Programming Microsoft® Robotics Studio

Sara Morgan

PUBLISHED BY
Microsoft Press
A Division of Microsoft Corporation
One Microsoft Way
Redmond, Washington 98052-6399

Library of Congress Control Number: 2008920201

Printed and bound in the United States of America.

1 2 3 4 5 6 7 8 9 QWT 3 2 1 0 9 8

Distributed in Canada by H.B. Fenn and Company Ltd.

A CIP catalogue record for this book is available from the British Library.

Microsoft Press books are available through booksellers and distributors worldwide. For further information about international editions, contact your local Microsoft Corporation office or contact Microsoft Press International directly at fax (425) 936-7329. Visit our Web site at www.microsoft.com/mspress. Send comments to mspinput@microsoft.com.

Acquisitions Editor: Ben Ryan
Developmental Editor: Devon Musgrave
Project Editor: Valerie Woolley
Editorial Production: nSight, Inc.
Technical Reviewer: Rozanne Murphy Whalen; Technical Review services provided by Content Master, a member of CM Group, Ltd.

Body Part No. X14-55481

For Steve Grand

Contents at a Glance

Table of Contents

Foreword

I am pleased to have an opportunity to provide a few words toward the introduction of *Programming Microsoft Robotics Studio*. I consider its creation a good indicator of the continuing interest in robot application development and, in particular, with Microsoft Robotics Studio.

Robots seem to captivate peoples' imaginations. They have been an enduring theme in science fiction and fantasy. And yet, they still seem somewhat removed from reality. In fact, over the past 20 to 30 years, robots have played an important part in facilitating industrial production and material handling. Few automobiles are produced today without robots playing some part in the process. They excel at doing dull, repetitive tasks and those that may be hazardous to humans. As different as the mainframe computers of the 1970s are from the personal computers we use today, industrial robots are not how we imagine interacting with robots. Most of us would prefer to see robots that are close in touch with us, operating safely in our workplaces and homes.

The good news is that, in recent years, we can see the beginning of this evolution–the progression of robots moving out of our factories and into our offices and homes. The first wave, though barely perceptible to us, has arrived in the form of smart appliances and technologies that make our lives easier or safer. For example, when I first learned to drive, I was taught that on icy roads I should gently pump a car's brakes to maintain control when my wheels began to slip. Today, with anti-lock braking–a form of robotic technology that senses the wheel slip and modulates braking–my car is able to do this by itself and do it more effectively. Likewise, my microwave oven has automatic settings for cooking food. I can put in a bag of popcorn, hit the "Popcorn" button, and the microwave adjusts the cooking time based on the steam generated by the popping corn.

We have also started to see the next wave of robots entering our homes, in the form of robot toys and simple single-task robots. When I was young, I built robots that I had to pretend were automated; my kids have grown up using LEGO Mindstorms, a true programmable robotic kit. And the number of sophisticated robot toys appearing on the market is growing. The WowWee RoboSapien is said to have sold over 3 million units in just 2 years, encouraging WooWee to produce an ever-increasing variety of other robots.

Not only is the number of toy robots on the market growing, they are also significantly advancing in terms of complexity, yet often selling at increasingly cheaper prices. LEGO Mindstorms began with a simple 8-bit processor brick programmed via IR; now LEGO sells a new 32-bit based NXT product that supports BlueTooth. This year's hit at the Japanese IREX robotic exhibition was the diminutive, but impressive, Tomy i-Sobot, a robot configured from 17 small servo motors, selling for $300. Just three years ago, it probably would have cost $3,000. The even more impressive, but ill-fated, Sony Aibo robot dog (discontinued last year to the dismay of many robot hobbyists) first sold for $3,000 and exited the market at $1,800. It has been replaced, in some sense, by the Ugobe Pleo dinosaur robot, available for $350.

On the more practical side, a great example of success is iRobot and its popular Roomba vacuum cleaner robot. Now in its third or fourth generation, Roomba has become the most recognized example of consumer robots. The company has added to its line of simple cleaning robots with Scooba for floor washing, Dirt Dog for shop floor clean-up, Verro for pool cleaning, and the recently announced Looj for roof gutter cleaning.

While these second-generation robots still may not fulfill our dreams as reflected in films and fiction, they should not be dismissed. The first personal computers also seemed primitive by today's PC standards and were merely toys or entertainment-based. The MITS Altair, typically recognized as one of the first PCs, bears little resemblance to today's desktops and laptops. Its processing power, memory, and programming has little in common with today's powerful machines. Computers like the Commodore Pet, the TRS-80 Model 1, and the popular Apple II had strange keyboards, limited memory, and initially saved programs to cassette tape. But these early products were indicators of where things were headed. Look at what a pervasive part of our work and home lives they have become. So, too, today's robot toys and simple-function robots foreshadow the technology that is to come.

Already the next generation of personal robots is starting to appear. In South Korea, companies like Yujin, Hanool, Grandport, MostiTech, and Microrobot, to name just a few, are beginning to introduce robots that are basically PCs on wheels, with equivalent processing, memory, storage, and network capabilities, at affordable prices. But they also feature proximity sensors for obstacle avoidance, navigational sensors, self-recharging base stations, and even social expression displays that give them the ability to reflect their condition in a natural, recognizable way.

While these PC-based robots are emerging, the generation just beyond them is starting to show up: robots that not only have all the capabilities just mentioned, but with dexterous manipulation arms, as well. While such robots are still priced outside the range of a consumer product, they demonstrate where the technology is headed and how quickly things are advancing. Robots with hands and arms represent almost unlimited potential when combined with advancements in vision recognition, navigation, and other critical technologies that personal robots will require.

This rapid evolution and advancement is very similar to what fueled the PC evolution and the advanced adoption of the Web as a key part of our lives. And yet, it has even greater potential, since the emerging generations of robots can build on the desktop and the Web, rather than starting from scratch as their PC cousins had to do. It is also fueled by intellectual and financial investments that are being made at an unprecedented rate. All major world economies are pouring money into robot research and startups to stimulate the industry's development. They all see robotics as a key player in future economic growth. Couple this with the technological advancements coming out of research and the academic world, as exemplified by the DARPA Urban Challenge, and the pump is primed for robot development to accelerate exponentially in the next 5 to 10 years.

Which brings us back to this book. As with the PC industry, success was not dependent simply on hardware advancements. It was software that really fueled the growth of the PC industry, the software that provided the reason for anyone to own the hardware. It is also for this reason that Microsoft has entered the robotics market. The Microsoft robotics initiative can be directly traced back to invitations from many of the early leaders of the robotics community to get involved, apply its assets, in a way that would provide greater access to development of the software applications and services that would be needed to drive this market. In response, Microsoft developed its robotics toolkit to help provide a catalyst for developing the software we believe can enable this emerging new market for robotics to bootstrap itself. We invested some of our best technology assets. The runtime programming model facilitated by CCR and DSS was actually designed to support software development for multicore processors and distributed processing scenarios. As such, our robotics toolkit is the first to harness and deliver this capability, not only to the robotics community but even for other types of applications beyond robotics.

I am pleased to have had the opportunity to introduce Microsoft's robotics toolkit. I consider it one of my greatest achievements in my 26-year history at Microsoft. It may also have the greatest impact, even beyond the first two releases of Windows or the many other products I have had the opportunity to introduce.

With *Programming Microsoft Robotics Studio* and Microsoft's robotics toolkit, you too have the opportunity to participate in what I consider to be the latest and most exciting evolution in PC technology. Whether you just enjoying tinkering with robots or intend to develop software that will deliver compelling value for robots, I hope you find this book helpful in getting you started.

Tandy Trower
General Manager, Microsoft Robotics Group

Acknowledgments

- To Steve Grand, for creating all of the graphics, reviewing my chapters, and convincing me that I could actually write this thing.
- To the super MSRS team at Microsoft that diligently reviewed every chapter and provided invaluable feedback. Specifically, thanks to George Chrysanthakopoulos, Henrik Nielsen, Joseph Fernando, Dave Lee, Kyle Johns, Andreas Ulbrich, and Tandy Trower.
- To the wonderful roboticists featured in the sidebars. I am lucky to know such a great group of talented and inspirational people. Specifically, thanks to Ben Axelrod, Roger Arrick, Raul Moreno, and Brian Cross.
- To Ben Ryan and Ken Jones from Microsoft Press for allowing me to write this book.
- To Valerie Woolley, who is by far my favorite project editor.
- To my three kids, who had to continuously hear, "Don't play with Mom's robots."

Introduction

Even though the field of robotics has been around for more than 50 years, it has only recently become more mainstream. Articles are now routinely featured in large online and print publications, such as MSNBC.com, CNET News.com, Forbes.com, *Wired Magazine*, and the *New York Times*. Most of these articles predict that robots are poised to enter the consumer marketplace in force. Take this quote from an article that appeared recently in the *Christian Science Monitor*:

> *Within a decade, observers say, the average American household will include one or two simple robots. And though they may not look like the ones imagined in science fiction, these robots – some available now – will play pervasive roles in the lives of regular consumers, says Lee Gutkind, author of* Almost Human: Making Robots Think.

For many years, robotics has been the exclusive domain of prestigious universities and industrial manufacturers. But now, more people are realizing that robotics offer tremendous advantages toward enhancing the quality of human life. In addition to companion robots, there are robotic dogs for the blind, household cleaning robots, and even robots that help children with learning difficulties such as autism.

About this Book

In an effort to make the field of robotics accessible to a larger group of developers, Microsoft has released Microsoft Robotics Studio (MSRS). MSRS is based on the popular .NET Framework. It offers developers a standard and consistent way to develop applications for an almost limitless variety of robotic platforms. Currently, version 1.5 of MSRS is available for download from the MSRS Web site (*http://www.microsoft.com/robotics*).

MSRS, along with four affordable and easy-to-work-with robots, will be featured in this book. The robots used in this book include the following:

- **Boe-Bot by Parallax** This three-wheeled robot is small, but it offers valuable experience for those new to electronic kits. In addition to the Boe-Bot kit, readers wanting to use this robot will need to purchase the eb500 Bluetooth module. The configuration steps required make the Boe-Bot the most time-consuming robot to assemble and configure.
- **Create by iRobot** This pre-assembled robot was built with developers in mind. It is based on the highly popular Roomba vacuum, but this model does not vacuum floors. Instead, you have the convenience of a pre-assembled and sturdy robot. Additionally, the payload capacity and expandability of this robot allows it to be used for several useful purposes.

- **Mindstorms NXT by LEGO** This surprisingly sturdy and useful robot is small, expandable, and easy to work with. The robot comes with five sensors, which can be easily attached to the base. Even though the robot is made with small LEGO bricks, this is not to be confused with a child's toy.
- **ARobot by Arrick Robotics** This three-wheeled robot features a large metal chassis, which makes it easy to expand. The robot was included as a means of demonstrating how new hardware can be supported (this is done in Chapter 7). In chapter 8, the ARobot is expanded to include an onboard laptop, which allows it to capture images using a USB Web camera.

One of the things that make this book particularly special is that all the chapters were reviewed by members of the MSRS product group team. The reviews were meant to ensure content accuracy, and, thus, I feel confident that the material covered is beneficial to the reader.

Who Should Read this Book?

This book was designed for people with limited or no experience with robotics—especially with MSRS. However, readers do need to be familiar with basic programming concepts. Those readers already experienced with Visual Studio will have the easiest time working with MSRS. This is especially true for readers already familiar with producing and consuming Web services.

This book may also be useful for developers experienced with MSRS. The last three chapters deal with advanced concepts that are covered only briefly, if at all, by the MSRS documentation.

Understanding the Terminology

If you are new to the world of robotics, you might not recognize some of the terms used throughout this book. This book assumes that you have some computer and programming experience, so I am not going to cover terms that apply to programming in general. The first time a term is mentioned in the book, a special reader aid box (such as the one below) will appear near the text that contains that term.

> **Actuators** In the world of robotics, an actuator is a physical device that can be used to move the robot. For example, the wheel, leg, or arm of a robot would qualify as an actuator. The motor of the robot can also serve as an actuator because this is typically used to power the robot. If you are a software developer, you can think of an actuator as a form of output such as a dialog box or a text box.

Additionally, readers new to the world of robotics may find it helpful to read through the glossary to find terms specific to robotics programming.

Chapter and Appendix Overview

This section will briefly cover what is included in each of the nine chapters. For readers unfamiliar with MSRS, it is best to start with Chapter 1 and work through the chapters one by one. These readers might also benefit from reviewing the documentation included with MSRS, which includes a set of step-by-step tutorials. Readers already familiar with MSRS might choose to skip to one of the more advanced chapters, such as Chapter 7, 8, or 9.

Chapter 1: Overview of Robotics and Microsoft Robotics Studio

This chapter begins with an overview of the field of robotics. This covers some of the fundamental challenges currently limiting progress in the field. The chapter then introduces MSRS, which seeks to address some of these challenges. The chapter covers the key components of MSRS, which include the Concurrency and Coordination Runtime (CCR) and Decentralized Software Services (DSS). It also covers the Visual Simulation Environment (VSE), which you can use to simulate robotics applications, and the Visual Programming Language (VPL), which provides a graphical tool you can use to easily build robotics applications using drag-and-drop techniques.

The chapter moves on to describe what is involved with a typical robotics application, as well as list the robots currently supported by MSRS. It concludes with a brief discussion of what readers can do if they want to work with unsupported hardware or execute code directly on the robot.

Chapter 2: Understanding Services

A service, which you can use to manage and control a robot's sensors or actuators, is a key component for an MSRS robotics application. Chapter 2 goes over, in detail, the various aspects of a service. This includes the most important component of a service: the state. The state is used to represent the service, and there are various ways you can manage the state. The reader also learns how to send and receive messages, which allows other services to read and update the state of a service. This is important because a typical robotics application involves the orchestration of more than one service.

Chapter 2 covers handling subscriptions, which involves configuring both the publisher and subscriber. The chapter also covers handling output through an Extensible Stylesheet Language Transformations (XSLT) file. This allows you to create a Web page that you can use to both view and update service state. The chapter concludes with a section on handling faults and debugging using debug and trace messages.

Chapter 3: Visual Programming Language

VPL is a graphical tool that lets developers build robotics applications by dragging and dropping activities onto a design surface. Activities, which are used to accept an input message and produce an output message, can include things such as variables, calculations, and conditional logic. Additionally, services can be inserted into an application as an activity.

This chapter features the Boe-Bot robot by Parallax. Readers step through the process of creating a VPL application for the Boe-Bot. The application simply responds to contact with the Boe-Bot's infrared detectors or whiskers by speaking a phrase. The remainder of the chapter covers topics such as debugging your VPL application, compiling the VPL application as a service, and creating a custom activity.

> **Note** Readers should keep in mind that VPL is a powerful tool and that this chapter does not cover all aspects of how you can use VPL. In reality, an entire book could be written about the subject of VPL alone. This chapter is simply meant to introduce VPL to readers new to MSRS.

Chapter 4: Simulation

Simulation provides a way for developers to work with robotics without purchasing and building a single robot. This chapter introduces the Visual Simulation Environment (VSE) tool and covers what you need to do to run a simulation. This includes the editor settings and information on how to work with entities. The entity is the core object in an MSRS simulation, and it can represent any object within the simulation scenario. Readers learn how to create new entities, based on the entity types provided with MSRS.

In addition to VSE, programmers can create simulations programmatically using Visual Studio. This chapter gives readers an overview of how this works and tells readers what references they need to add to the service project. The steps required to create new entities are briefly covered.

> **Note** As with Chapter 2, this chapter serves only as an overview of simulation and is not intended to cover all aspects related to simulation. The subject is simply too big to cover in one chapter and may need an entire book devoted to it. This chapter introduces simulation to those readers new to MSRS.

Chapter 5: Remote Control and Navigation

This is the first chapter that includes code samples on the book's companion Web site. The particular code project associated with this chapter is named BasicDrive. The BasicDrive service demonstrates techniques used to operate a robot using button controls on a Windows form. Additionally, the service demonstrates how a service can receive notifications from a robot's sensor.

This chapter introduces the Create robot by iRobot. This fully assembled robot is one of several supported by MSRS. This chapter steps the reader through all of the code used to create the BasicDrive service. This includes the steps to add a Windows form, as well as the code that is initiated when the user clicks a button on the robot. By using generic service contracts, you can use the BasicDrive service to operate any robot with a two-wheel differential drive system.

Chapter 6: Autonomous Roaming

This chapter uses two service projects (both located on the book's companion Web site) to demonstrate ways you can cause a robot to roam autonomously. This means that the robot should move around an area and be able to avoid obstacles without being directly programmed to do so.

We use the first version of the service, named Wander, to move a LEGO NXT robot around a room. The NXT robot will move away from an obstacle only after the front touch sensor has made contact with the obstacle. Although this version does cause the robot to roam autonomously, the method is reactive and does not always result in desirable behavior.

Version 2 represents a more proactive approach to autonomous roaming. In this version, the NXT sonar sensor is used to detect objects before contact is made. This not only results in less collisions but also helps prevent the robot from getting stuck.

Chapter 7: Creating a New Hardware Interface

MSRS supports several pre-assembled robots and robot kits. However, this does not mean you cannot use MSRS with an unsupported robot or custom-made hardware. This chapter covers the steps needed to create your own hardware interface. The sample code referenced in this chapter and included on the book's companion Web site is used to interface with a robot from Arrick Robotics named the ARobot.

Creating a new hardware interface involves creating several layers. The first layer is the onboard interface, and this is the part that executes directly on the robot hardware. The next layer is the control or brick service. This part will communicate directly with the onboard interface to send commands to the robot and receive data from the robot's sensors. Another layer consists of multiple services that represent each of the robot's sensors or actuators. The last layer is a test service that can be written as a traditional service using Visual Studio or using the VPL tool.

Note This chapter represents one of the advanced chapters in this book and may be difficult for readers new to MSRS.

Chapter 8: Building a Security Monitor

The final code-based chapter steps the reader through building a security monitor robot. Using the new hardware interface created in the previous chapter, this chapter uses the ARobot as the hardware for the security monitor. The SecurityMonitor service uses an ordinary USB Web camera, which must be attached to an onboard laptop. Because the ARobot does not include a laptop, a special support was added to the ARobot's metal chassis.

Images are captured by the Web camera every few milliseconds. Each image is compared with the preceding image to determine if movement was detected. If motion is detected, then an e-mail notification is sent to a predetermined e-mail address, and the motion detection is temporarily disabled. The e-mail recipient can then use an Internet connection to view the image in which movement was detected. The recipient can also use the Web page to re-enable motion detection and change the e-mail address that notifications are sent to.

Note This chapter represents one of the advanced chapters in this book and may be difficult for readers new to MSRS.

Chapter 9: Future of Robotics

This chapter does not include grand predictions about where the field of robotics will be in 50 or 100 years. Instead, it focuses on solid information based on work already done or currently in progress. The first area covered is the future of MSRS. Based on responses received from the MSRS team, this section reveals a future direction for the maturing product.

The chapter also features a section on potential applications. This section covers the following areas of robotics applications: smart appliances, caring for the elderly, performing dull or dangerous jobs, performing exploration jobs, and providing specialized assistance, companionship, and entertainment. The chapter concludes with a brief look at how artificial intelligence techniques could be incorporated into MSRS applications to make them more useful.

Appendix A: A Brief History of Artificial Intelligence

This appendix includes a brief overview of the field of artificial intelligence (AI). This includes a listing of a few popular branches of AI and a review of the current state of AI. This appendix is meant to provide background for those readers new to the field of robotics and AI.

Appendix B: Configuring Hardware

This appendix includes specific setup instructions for working with each of the robots featured in this book. This includes the Boe-Bot by Parallax, the Create by iRobot, and the NXT Mindstorms by LEGO.

Hardware and Software Requirements

This section lists the hardware and software requirements needed to work with MSRS and the sample code provided with this book.

Hardware Requirements

It is recommended that you use a development machine such as a laptop or desktop computer. The minimum requirements for this development machine are as follows:

- Personal computer with 600 MHz Pentium III–compatible processor (1 GHz or more recommended)
- 512 MB of RAM (1GB or more recommended)
- 200 MB free hard disk space for the MSRS installation
- Up to 1.3 GB free hard disk space may be required when installing the Visual Studio Express Edition
- Super VGA (1024 X 768) or higher resolution video adapter and monitor
- Internal graphics card capable of supporting vertex shader VS_2_0 or higher and pixel shader PS_2_0 or higher (needed to run the simulations in Chapter 4)
- Keyboard and Microsoft mouse or compatible pointing device

Software Requirements

You will need to install the full version of MSRS, which is available for download from the Microsoft Web site. As long as you use MSRS to follow along with the examples in this book and do not use it for commercial distribution purposes, you may download and install it for free. You will also need a copy of Visual Studio installed on your development machine. If you do not already have a copy installed, you can download an Express Edition from the Microsoft Web site for free.

Installing Microsoft Robotics Studio 1.5

1. Browse to the MSRS download Web page using Internet Explorer. The download Web page is available through the following URL: *http://msdn2.microsoft.com/en-us/robotics/aa731520.aspx*.
2. Click the Microsoft Robotics Studio (1.5) link. When prompted, save the setup executable to an area on your local hard drive.
3. Click the Samples Update for Microsoft Robotics Studio (1.5) link and save the zip file to an area on your local hard drive.

4. Double-click the Microsoft Robotics Studio (1.5) setup launcher file to begin the installation. You need to complete this installation before you can install the samples update.
5. Wait while InstallShield wizard inspects your machine and extracts any setup files. If previous installations exist on the machine, the MSRS help file for the previous version will be uninstalled.
6. The installation will take several minutes to complete; the exact time will vary depending on the speed of your machine. The MSRS installation will install the correct version of the .NET Framework and the AGEIA PhysX engine, which is needed for the simulation environment.
7. After the MSRS installation is complete, unzip the samples update file to a folder on your hard drive. Double-click the application executable and follow the instructions to complete installation of the samples update.

Installing Visual Studio C# 2005 Express Edition

1. Browse to *http://www.microsoft.com/express/2005/download/default.aspx* using Internet Explorer and select a language from the Visual C# 2005 Express Edition download box.
2. Browse to a location on your hard drive in which to download the vcssetup application file.
3. Double-click the setup file and wait until the installation copies the required setup files to your hard drive. When it is complete, click Next to continue the installation.
4. Select the I Accept The Licensing Terms And Conditions check box and click Next to continue the installation.
5. Optionally, you can choose to install the MSDN 2005 Express Edition. Click Next to continue the installation.
6. Click Install to begin the installation, which will take several minutes to complete. The exact time will vary depending on the speed of your development machine.
7. Return to the download page (*http://www.microsoft.com/express/2005/download/default.aspx*) and select the Download Visual C# 2005 Express SP1 link.
8. Copy the setup executable to an area on your local hard drive.
9. Double-click the executable and follow the instructions to complete installation of the service pack.

How to Access Code Samples

This book features a companion Web site that makes available to you all the code used in the book. This code is organized by chapter, and you can download it from the companion site at this address: *http://www.microsoft.com/mspress/companion/9780735624320*.

All of the code samples provided on the book's companion Web site are written in C#. They are accessible by using Visual Studio 2005 to open the associated solution file. Readers can use any version of Visual Studio 2005 to access these files, including the Visual Studio Express Edition, which is available as a free download from the Microsoft Web site. To execute the code samples, readers will need to assemble and configure the associated robotics hardware.

Readers unable to access the hardware from each chapter can either modify the code to work with their particular hardware platform or follow along with the code without attempting to execute the application. The code samples available on the book's companion Web site include the following:

- **Chapter 5** BasicDrive/BasicDrive.sln–Loads the BasicDrive.csproj project in Visual Studio. To execute this application, readers will need to configure the Create robot by iRobot (see the section in Appendix B titled "Configuring the iRobot Create").
- **Chapter 6** Version1/Wander.sln–The first version of this application will use the touch sensor from the LEGO NXT robot as a contact sensor.

  Version2/Wander.sln–Version 2 of the Wander service will utilize the sonar and sound sensors from the LEGO NXT robot. Before executing this service, the reader will need to assemble and configure the LEGO NXT robot (see the section in Appendix B titled "Configuring the LEGO NXT").
- **Chapter 7** ARobot/ARobot/ARobot/ARobot.sln–Opening this solution file with Visual Studio will load multiple projects. The first project, named ARobot, represents the control or brick service. The second project, named ARobotServices, holds all the services used to control the ARobot's sensors and actuators. The last project, named ARobot-Test, should be the Startup project (this is the case if the text for the project is bolded). Before executing the ARobotTest service, the reader will need to assemble and configure the ARobot by Arrick Robotics (see the section in Appendix B titled "Configuring the ARobot").
- **Chapter 8** SecurityMonitor.sln–Loads the SecurityMonitor.csproj project in Visual Studio. To execute this application, you will need to assemble and configure the ARobot. The services used to communicate with the ARobot would have been created in Chapter 7. You will need to compile that project on your machine before you can start working with this project.

Note Even though code samples are not provided for the Boe-Bot robot, Appendix B includes a section titled "Configuring the Boe-Bot." This robot is introduced in Chapter 3.

Find Additional Content Online

As new or updated material becomes available the completments your book, it will be posted online on the Microsoft Press Online Developer Tools Web site. The type of material you might find includes updates to book content, articles, links to companion content, errata, sample chapters, and more. The Web site will be available soon at *http://www.microsoft.com/learning/books/online/developer* and will be updated periodically.

Support for This Book

Microsoft Press provides support for books and companion content at the following Web site: *http://www.microsoft.com/learning/support/books/*.

Questions and Comments

If you have comments, questions, or ideas regarding the book or the companion content, or if you have questions that are not answered by visiting the sites previously listed, please send them to Microsoft Press via e-mail to

mspinput@microsoft.com

Or via postal mail to

Microsoft Press
Attn: Programming Microsoft Robotics Studio *Editor*
One Microsoft Way
Redmond, WA 98052-6399

Please note that Microsoft software product support is not offered through the above addresses.

Chapter 1

Overview of Robotics and Microsoft Robotics Studio

When you think of robotics, you typically call to mind images from television or the movies. Depending on your age, you might think of C-3PO or R2-D2 from the famed *Star Wars* series. Or you might think of the android Data in the *Star Trek: The Next Generation* television series. More than likely you don't think of robots assuming a practical or realistic role in your own life—but that may be about to change.

If Bill Gates has anything to say about it, there will be a robot in every home in the not-too-distant future. In an article published in the January 2007 issue of *Scientific American*, Gates predicts that the robotics industry is about to break open in much the same way the personal computer industry did in the 1980s. Gates writes:

> *I can envision a future in which robotic devices will become a nearly ubiquitous part of our day-to-day lives. I believe that technologies such as distributed computing, voice and visual recognition, and wireless broadband connectivity will open the door to a new generation of autonomous devices that enable computers to perform tasks in the physical world on our behalf.*

In this chapter I'll give you an overview of the robotics field, covering some of its history and some of its challenges, and I'll introduce Microsoft Robotics Studio (MSRS), a robotics programming platform developed by Microsoft that will help make Gates's predictions a reality.

Current World of Robotics

Robotics is a field that has been around for several years. It has long been associated with artificial intelligence, but, for many people, robotics is seen as a practical solution in a world desperate for automation.

> **Note** For readers curious about the world of artificial intelligence, see Appendix A, "Brief History of Artificial Intelligence."

For several years, robotics has been widely used by industrial companies to perform jobs that were considered tedious or that were physically beyond the human capabilities. For example, robots have long been used on motor vehicle assembly lines to streamline the manufacturing of new motor vehicles. In the motor vehicle industry, one robot is used for every 10 workers. According to the International Federation of Robotics (*http://www.ifr.org*), the world market for industrial robots peaked in 2005 with a 30 percent increase over the previous year.

Beyond industrial robots is the newly emerging market of service robots. These robots are available for both personal and professional use. You may already be familiar with the floor-cleaning robot known as Roomba. iRobot (*http://www.irobot.com*), the maker of the Roomba, manufactures a variety of practical robots that do everything from clean your floors and pools to disable bombs. Roomba sales have already surpassed 2 million units, and you can purchase one in stores across the United States.

Service robots for professional use include those used for underwater exploration, defense and security, construction and demolition, and medical purposes. In the area of robots for personal and domestic use, vacuum and lawn-mowing robots account for the highest-selling products. This number is expected to change as robots are now being built for handicap assistance, personal transportation, and home security. The market for these types of robots is expected to increase significantly over the next few years.

Another quickly emerging market is entertainment robots. This includes not only robots used as toys and companions but also robots used by educators and hobbyists. WowWee (*http://www.wowwee.com*), maker of the Robosapien, shown in Figure 1-1, offers a line of robotics toys. The Robosapien V2, which claims to be a fusion of personality and technology, comes with a remote control and 100 preprogrammed functions, such as walking, grunting, burping, and dancing. The latest version is even capable of autonomous "free roam" behavior.

Figure 1-1 Robosapien is a popular robotics toy made by WowWee that is capable of autonomous behavior.

According to projections by the International Federation of Robotics, the number of service robots for personal use is expected to reach 5.6 million units for the period 2006 to 2009. The market for these domestic robots will continue to grow, and there will soon be an urgent need for developers with knowledge of robotics programming.

Challenges in Robotics

Currently the greatest challenges facing robotics researchers have to do with operating robots in dynamic environments and dealing with an infinite number of possibilities. If you were to break down your movements into a series of specific tasks, the number of tasks would approach hundreds or thousands, depending on the number of obstacles in your room.

For example, to move a robot across a room full of obstacles, the robot must be able to avoid all the obstacles before arriving at the destination. This is hard enough when the obstacles are constant things like chairs and tables, but what if a person steps directly in front of the robots path? To avoid hitting the person, the robot will need to react quickly, and this is not always an easy thing for a robot to do. The robot would need to constantly collect data from its sensors. The data collected from these sensors would be used to indicate whether an object was in the robot's path. But just collecting the data is not enough. You need to be able to process the sensory data and then instruct the robot to do something, such as turn to the right or stop altogether. As humans, we take this kind of simple decision-making for granted because it comes so naturally to us. But, for robots, these simple things can take an enormous amount of processing power. Each step or movement must be programmed specifically because the robot is not able to think for itself.

Despite the complexity involved with getting a robot to perform human-like tasks, robots can do some things quite easily. Wheeled robots, similar to the Roomba vacuum, can easily navigate a room and avoid obstacles by using a series of sensors. Crawling robots, such as those that mimic insects, are very successful at navigating rough terrain with minimal intelligence. Despite that, challenges remain, and you as a developer can start building useful robotic applications using the resources available today.

Robotics at Microsoft

A few years ago, Tandy Trower, a long-standing fixture at Microsoft, met with many people in the robotics industry to determine what their greatest needs were. Trower met with robotics researchers both in academia and in the commercial industry. He also considered the needs of hobbyists and students. His research resulted in a wish list, similar to the following:

- Easily configure sensors and actuators and be able to run them asynchronously
- Start and stop software components dynamically
- Monitor the robot interactively and as it is operating
- Allow more than one person to access a single robot or allow one person to access multiple robots
- Reuse software components across robots

The needs Trower identified were simple and fell into some well-covered robotics territory: they involved the basic requirements of operating and monitoring a robot across multiple platforms. The common problem was that each time someone built a robot, the creator used different hardware to build it and different software to operate it. Also, building high-level robots was expensive, and this often meant researchers had to share access to a single robot.

Trower felt that if he could create a platform that allowed robotics researchers to work with robots in a standard and consistent way, he could open the doors of robotics to more people. He also thought that the creation of such a platform, which resolved many of the day-to-day details associated with robotics, would allow researchers to focus on the harder-to-solve problems.

Introducing Microsoft Robotics Studio

In June 2006, Trower and a small group of developers he assembled announced the availability of MSRS. MSRS was the direct result of the interviews Trower had done with the robotics community. It was the platform he hoped would solve many of the obstacles facing robotics researchers and potential hobbyists.

MSRS is a Microsoft Windows–based environment that offers a service-oriented runtime, visual authoring tools, tutorials, and documentation. The MSRS toolkit provides a platform and tools that can be used by both commercial and academic robotics researchers. MSRS also offers a visual programming language that can be used by students and hobbyists. Since Version 1.0 was released in late 2006, the robotics group at Microsoft has released one major version of MSRS. Version 1.5 improves on the existing product offering by including significant enhancements to the visual simulation environment and visual programming language.

Even though the development group creating MSRS was small (only 11 developers), they were able to get a release together very quickly. This was partly because MSRS is based on the Concurrency and Coordination Runtime (CCR). The CCR is a .NET library designed to make asynchronous programming much easier than it was previously. Coincidently, it was being developed around the same time that Trower was investigating ways to make robotics easier. The MSRS team that Trower put together consisted of two developers from the CCR development team. They were able to leverage the work done by the CCR group to deliver MSRS as quickly as possible.

The latest release of MSRS, version 1.5 (available for download at *http://www.microsoft.com/downloads*) included several enhancements that made the product much easier to work with. MSRS is a work in progress, and the team has been hard at work improving the tools and services that come with MSRS. They are working with several third-party hardware vendors to integrate MSRS with existing robot hardware.

Microsoft Robotics Studio Licensing Options

MSRS comes with two forms of licensing: commercial and noncommercial. If you are a student, hobbyist, or someone looking to learn more about robotics, you would need only the noncommercial license. This means you will not attempt to make money using any of the robots you work with. If this is all you require, MSRS is free to download (available at *http://msdn2.microsoft.com/robotics/default*) and use.

If you are a commercial developer and you want to use your robot in a way that will generate revenue, you can purchase a commercial license for $399 (U.S. dollars). The commercial license allows you to distribute up to 200 copies of any runtime components you create with the software.

Selecting a Development Language

Because MSRS runs on the .NET Framework and you will use Microsoft Visual Studio to design your robotics application, you can use any of the languages supported by the .NET Framework. Most of the samples and tutorials available with MSRS are written with C# or Visual Basic .NET. The sample code used throughout this book will be written in C#, but this does not mean you cannot use Visual Basic .NET if that is the language you prefer.

MSRS also allows you to use the Python scripting language, which is an extensible and object-oriented language distributed under an open-source license. Microsoft offers a version of Python that works with .NET called Iron Python. MSRS includes sample code written with Iron Python for those programmers who prefer that language. For more information on Python, visit *http://www.python.org/*.

MSRS introduces a graphical programming language called Visual Programming Language (VPL). VPL offers beginning programmers and students a visual interface that allows them to use drag-and-drop techniques to build a robotics program. VPL can also be useful for experienced programmers who want to quickly build an application prototype. Programs written in VPL can be converted easily into C# applications. Therefore, an application that began in VPL can be extended and enhanced using Visual Studio and the .NET Framework.

Key Components of Microsoft Robotics Studio

MSRS contains services and tools that allow you to design, deploy, and debug your robotics applications. Applications built with MSRS will run over the Web or a local intranet as a collection of state-bound services that are isolated from each other. This will allow developers to build Web-based robotics applications that are lightweight and flexible. This section will provide an overview of the key components that make up MSRS.

Runtime

The MSRS runtime, as depicted in Figure 1-2, makes use of three lower-level runtimes: Concurrency and Coordination Runtime (CCR), Decentralized Software Services (DSS), and the .NET Common Language Runtime (CLR) 2.0. CCR is a message-oriented model that allows applications to coordinate asynchronous processes and exploit concurrency in a very efficient way. DSS is a lightweight, service-oriented runtime that combines the principles used to power the World Wide Web with the architecture used to design Web services. CLR supports both base runtimes and provides access to the .NET Framework.

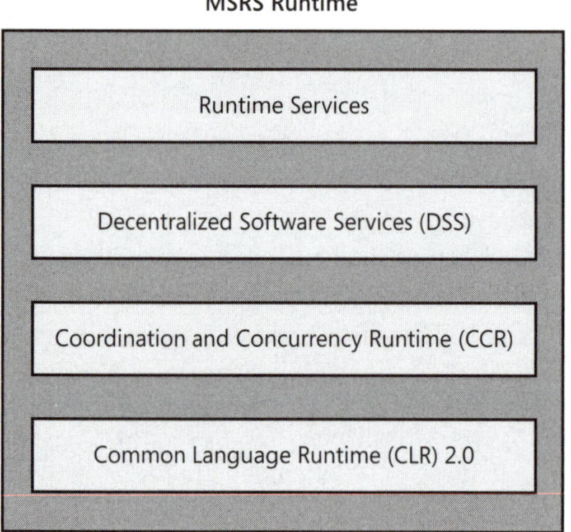

Figure 1-2 The MSRS runtime is built on top of CLR 2.0 and uses two key technologies—CCR and DSS—to build a set of services that supports the central object in MSRS, the service.

Concurrency and Coordination Runtime

The CCR is the piece that ties the MSRS runtime to a service. Introduced at the same time as MSRS, the CCR allows applications to perform asynchronous operations. The CCR is a .NET library that can be used in any application that requires asynchronous processing. It shields developers from all the difficulties involved with writing their own threading code.

You might ask why asynchronous processing is so important for robotics applications. It's important because, without asynchronous processing, your robot could move only one wheel at a time or one arm or leg at a time. Without asynchronous processing, the robot would have to wait for each service to complete what it is doing before it can start another service.

Robots are typically equipped with multiple *sensors* that read all sorts of data in their environments. At the same time, they have actuators that are used to perform a movement or accomplish a goal. All of these sensors and actuators must be coordinated to accomplish a common goal, which is operating the robot. The CCR manages the individual threads that are needed to make this happen.

Sensors Robots use sensors to gather input about their environments. Essentially, they allow your robot to see, hear, touch, and smell. The types of sensors your robot uses will vary; dozens are available. Most robots come with at least one type of contact sensor. This can include whiskers that are sensitive to touch or phototransistors that are sensitive to light. Beyond that, the robot could have a sensor to record temperature or a GPS receiver to identify its location.

Without the CCR, developers would have to create several callback methods to coordinate services. Applications would also require complex threading code to prevent blocking of asynchronous operations. This could quickly become cumbersome to manage. The developer is shielded from all of the work being done under the covers to manage asynchronous operations and deal with parallel execution. For more information about the CCR, refer to the CCR user guide available at *http://msdn2.microsoft.com/library/bb483107*.

Decentralized Software Services

DSS is a lightweight application model that allows developers to monitor services interactively in real time. DSS is based on the Representational State Transfer (REST) design principles. REST encompasses a set of principles used to define the technologies that power the World Wide Web. The Web is a stateless environment, which means that every time you access a new Web page, the information returned from the last request is discarded. Each request is independent of the last. Although .NET Web developers can use server-based containers such as session variables to store information between requests, each request is independent of any other.

REST is not a standard, but it does use standards to accomplish its goals. The REST principles specify how resources are used, and the requested information is transmitted across the Web. It enables the Web to be stateless, layered, cacheable, and thus highly scaleable. These simple principles are what allow millions of people to successfully use the Internet each day.

To operate and monitor robots remotely, MSRS needed a distributed solution that operated efficiently. It made sense to use the REST principles because they provided an efficient way of traversing the Web. REST supported the widely used Hypertext Transfer Protocol (HTTP) communication protocol, and, along with Extensible Markup Language (XML), this could be used to return structured XML data quickly.

MSRS also needed a transport protocol that was flexible and able to accommodate distributed scenarios in a platform-independent manner. The transport protocol Simple Object Access Protocol (SOAP) was the best choice, but the MSRS developers needed something like SOAP, only with a few extra benefits. They needed to support subscriptions. Subscriptions allow a *service*, the core object used by MSRS, to receive event notifications.

> **Service** A service, which is accessible over the Web and local intranet with a Uniform Resource Identifier (URI), is a key object in an MSRS robotics application. A service is a set of code used to perform some operation on one or more objects. Services are like building blocks for a robotics application, and one or more are used to collect data from robots and then instruct them on what to do.

To support subscriptions, the MSRS designers incorporated principles from Web services. This allowed them to provide a flexible solution that took advantage of structured data manipulation and event notification. By using principles from the REST model along with Web services, MSRS is able to offer an extensible and efficient platform that supports access across different machines. When you combine this with the ability of DSS to isolate each service from one another and the ability to separate state from behavior, you have a powerful tool in the development of distributed robotics applications.

DSS also offers the ability to transport services through Transmission Control Protocol (TCP). This is sometimes advantageous because TCP provides minimal Central Processing Unit (CPU) utilization and reduced latency. At the same time, TCP provides an HTTP/XML path to every service. This gives you the ability to observe remotely the behavior of your robot through a Web browser.

> **Note** Just because MSRS uses Internet protocols as the basis for the runtime does not mean that your robotics applications will have to run across the Internet or even the local intranet. In fact, all the components of your robotics application can reside on the same machine. It also does not mean you have to become a Web developer and start designing Web sites. This section was meant to provide an understanding of the principles that underlie the base DSS runtime.

Decentralized Software Services Protocol MSRS offers a SOAP-based protocol named Decentralized Software Services Protocol (DSSP). DSSP, which was created specifically for MSRS, extends upon HTTP and offers an alternative method for message-based operations. For example, one of the HTTP operations allowed through MSRS is GET. This operation is used to return requested data as raw XML. Although MSRS supports the HTTP GET operation, in order for one service to handle the data sent from another service, it needs to be sent as a SOAP message.

One of the primary advantages to using services is that they can work together to accomplish a common task. It is often necessary for one service to pass information to another service. For this to be done, each service must accept incoming messages in a machine-readable format, such as a SOAP message.

DSSP offers comparable replacements for all the HTTP-based operations, such as GET, POST, PUT, and QUERY. DSSP also offers additional operations that specifically support functionality required by MSRS. For example, to support subscriptions, MSRS needs an operation that requests event notifications be sent to a subscribing service. HTTP does not support a SUBSCRIBE operation, but DSSP does. The SUBSCRIBE operation was added specifically to support this functionality in MSRS.

DSSP is not meant to be a replacement for HTTP or Web services; it is simply an extension of HTTP that takes advantages of some of the things offered by Web services. For more information about DSSP, refer to the specification document available at *http://download.microsoft.com/download/5/6/B/56B49917-65E8-494A-BB8C-3D49850DAAC1/DSSP.pdf*.

Runtime Services DSS provides an extensible foundation that allows it to offer a set of infrastructure services known as the runtime services. These services are used to manage other services created with MSRS. Services managing services might seem confusing, but that is exactly what is happening. For example, there is a runtime service that renders an HTML-based user interface known as the Control Panel. The Control Panel, which lists all the services running on your development machine, is used to locate and run other services.

Visual Simulation Environment

The Visual Simulation Environment (VSE) tool included with MSRS offers programmers a way to enter the world of robotics without any robotics hardware. Powered by the AGEIA PhysX engine, VSE renders advanced physics graphics comparable to graphics used in the best computer games on the market. VSE also uses the Microsoft DirectX 9 runtime components and Microsoft XNA (not acronymed) to do 3-D rendering. This means the machine you use to do simulations must support DirectX 9 pixel/vertex shader standards, but that is not a problem for most of today's mid- to high-end desktop machines.

VSE is great for students and hobbyists who want to learn about robotics and who do not have the time or money to invest in expensive robots. Typically, the expense of advanced hardware forces teams of students or commercial developers to schedule time to work with a single robot. With VSE, each team member can work with the simulation tool as long as required and then apply his or her individual work directly to the actual robot.

VSE can also be useful for developers wanting to prototype their robotics project. Instead of having to spend a lot of time building and configuring robots, programmers can test their code before they risk damaging expensive equipment. Obviously, the downside to using a simulation tool is that it is not entirely realistic. In the "real world," of course, robots encounter unforeseen obstacles and have to react in ways their programmers never expected. For this reason, simulation is useful at the beginning of a project, but only use of an actual robot will yield "real world" results.

The simulation engine, which in itself is just another service, uses a native physics engine wrapper and library that are built on top of the AGEIA PhysX engine. (See Figure 1-3.) Optionally, you can purchase the PhysX processor, which was created specifically to enhance the PC gaming experience. The PhysX processor will support hardware acceleration, which may be necessary when working with some heavily CPU-intensive simulations.

Tip You can still work with simulations without the add-in card; you will be limited to working with a smaller number of objects. If you need to do simulations involving several hundred robots performing at the same time, you will likely need to purchase the PhysX processor.

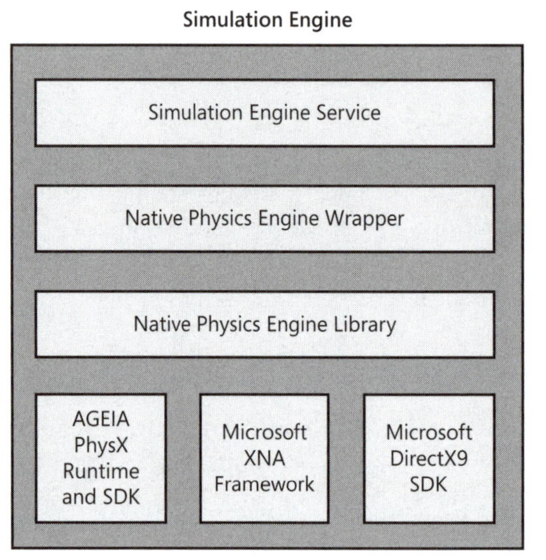

Figure 1-3 The VSE simulation engine is based on the AGEIA PhysX engine, along with Microsoft XNA Framework and the DirectX 9 runtime.

The simulation engine is also based on the Microsoft XNA Framework and DirectX 9 software development kit (SDK). The XNA Framework is a set of class libraries built against the .NET Framework that allow game developers to create games easily by using C#. The DirectX 9 runtime, which is used by the XNA Framework, allows developers to build cutting-edge multimedia applications such as VSE. If you do not already have these libraries on your development machine, the MSRS installation program will include them and tell you if you first need to uninstall an incompatible version of the DirectX 9 runtime.

VSE works by managing entities, which can represent any physical object, such as a robot, an obstacle, a camera, and even the sky. Services are used to move the robots around and gather data from sensors. Orchestration services are used to coordinate a group of services. You can use the same services in the simulated world as you do using the MSRS runtime. This makes VSE a useful tool for prototyping a robotics project.

Graphics scenes can be rendered in one of four modes: visual, wireframe, physics, or a combination of all three. Visual mode (see Figure 1-4) represents what the robot would look like in the real world. Wireframe mode is a pixilated version that shows just an outline of the objects. Physics mode is a simpler view that shows how objects are modeled by the physics engine. The simulation tool will be covered in detail in Chapter 4, "Simulation."

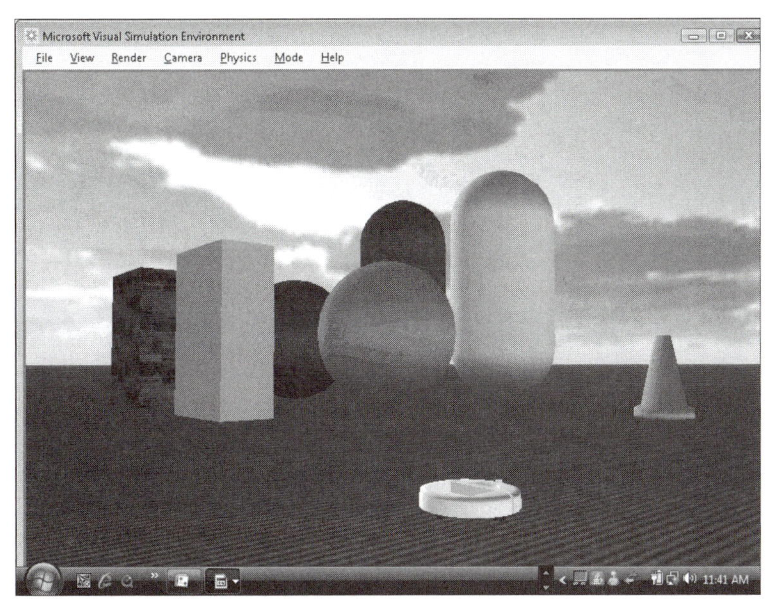

Figure 1-4 Visual-mode rendering of a simulation scene.

Visual Programming Language

MSRS includes the visual code-generation tool named VPL. Integrated with Visual Studio, VPL can be used by both beginning and experienced programmers. To build a program, a user only needs to drag and drop blocks, which represent activities or services, onto the design surface. VPL offers two kinds of activities. The first, basic activities, includes input data, calculations, and variables. The second, service activities, includes the built-in services that come with MSRS as well as the services that you build. VPL also includes the runtime services used to power MSRS.

Connections are used to specify that the output from one block will be used as the input for another block. Figure 1-5 displays a basic program that consists of two blocks with one connection. This program displays the phrase "Hello World" inside a dialog box. In this case, the "Hello World" string is entered into the first block and then sent as output to the second block. The second block receives it as input and then displays it as an alert.

VPL can be used to build complex programs such as those that collect and process data from multiple robot sensors. It can also be used to create autonomous behavior for any robot that has a distance-measuring device and *differential drive system*. Even experienced programmers may find that it's quicker to build an application using VPL and the built-in services provided with MSRS.

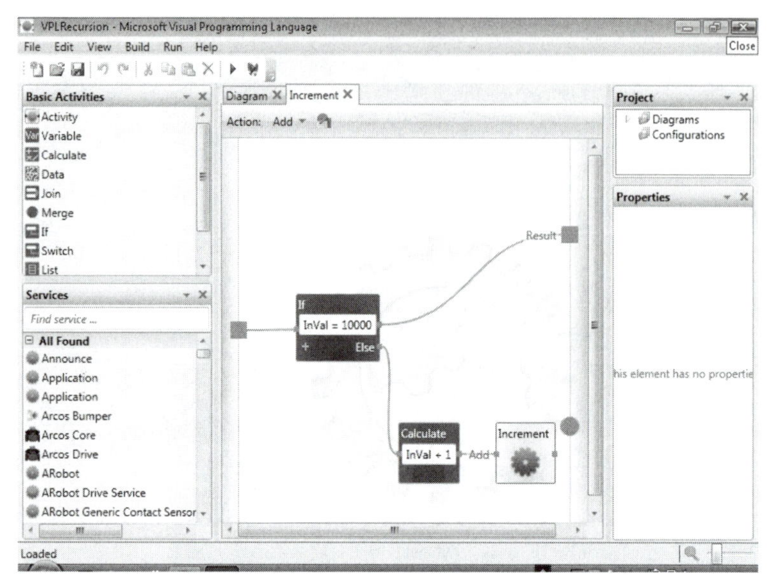

Figure 1-5 The VPL tool provides a visual application design environment in which users drag and drop blocks onto a design surface.

VPL can be used to build complex programs such as those that collect and process data from multiple robot sensors. It can also be used to create autonomous behavior for any robot that has a distance-measuring device and *differential drive system*. Even experienced programmers may find that it's quicker to build an application using VPL and the built-in services provided with MSRS.

> **Differential drive system** Many robots today (and especially those currently supported by MSRS) are wheeled robots. To operate a wheeled robot, you must utilize the robot's differential drive system. This is used to steer the robot in any direction. Each wheel on a robot is powered by a motor, and these motors accept values that indicate how much power to apply to each wheel. The direction that your robot moves is determined by the difference between the power sent to the right and left wheels. At this point, it is not important for you to know exactly what values to use, but understanding how this works is critical to operating most robots. And do not worry: MSRS comes with built-in services that will help you drive your robot.

You can save VPL applications as files with an .mvpl file extension. Several of the robotics tutorials that come with MSRS include VPL files. This will be particularly helpful for new and experienced programmers as they are learning how robotics programming works. Chapter 3, "Visual Programming Language," will provide more information about how VPL works.

What Is a Typical Robotics Application Like?

Most robotics applications involve a robot that collects data from one or more sensors and then responds accordingly. For example, you could create an application that operates a robot with a sound sensor. Some sound sensors can detect different decibel levels, which is just a measurement of sound pressure. The higher the decibel level, the louder or closer the source of the sound. You could create an MSRS program that causes a robot to move or perform some action every time someone claps his or her hands or shouts a command. The robot would not actually understand the command because the sound sensor only detects a level of sound. But, if the sound fell within a certain decibel range, you could assume it was an intentional command and have the robot respond.

MSRS programs can be as simple or complex as they need to be. In fact, they do not always have to involve robotics hardware. You can create a service that simply processes input and produces some output. In the next chapter, you will see an example of a service that performs like the classic Hello World example. In this service, the input is the phrase "Hello World." The service will simply return the phrase within an XML message that will be displayed in a Web browser.

How complex the robotics application becomes will depend on how much time and effort the programmer wants to devote to the project. Programmers already familiar with creating .NET programs can install MSRS and immediately create a program by following the documentation for Service Tutorial 1. (This documentation can be found online at *http://msdn2.microsoft.com/library/bb483064*.) Within an hour they could have their first robotics application up and running. The application would not include an actual robot because the Service Tutorials do not involve any robotics hardware.

The amount of time required to write a robotics application depends on several things, such as the complexity of the robot you have selected, the number of robots you are using, and the functionality required of the robot. For example, if you want to program one robot that simply moves forward when it receives a command, this will take significantly less time than programming a group of robots to play soccer.

Listing of Supported Robots

Three robots supported by MSRS will be used in examples throughout this book. They are the Create by iRobot (see Figure 1-6), the Boe-Bot by Parallax (see Figure 1-7), and the LEGO Mindstorms NXT (see Figure 1-8). The Create robot is an affordable robot that requires no assembly. The Boe-Bot is also an affordable robot that requires only a little assembly and offers a wide range of possibilities for expansion. The Mindstorms NXT is a surprisingly powerful and affordable robot that includes several useful sensors as well as built-in Bluetooth capabilities.

All three robots featured in this book were chosen because they are small, affordable, and easy to work with. This does not mean that you cannot use MSRS with industrial-strength robots or with complex scenarios involving custom hardware. In fact, there is a group of Princeton students using MSRS to compete in this year's annual Defense Advanced Research Project Agency (DARPA) Grand Challenge. These undergraduates have designed 25 services running on five different servers to qualify for the autonomous vehicle challenge. (For more information, see the section titled "Gearing up for the Urban Challenge.")

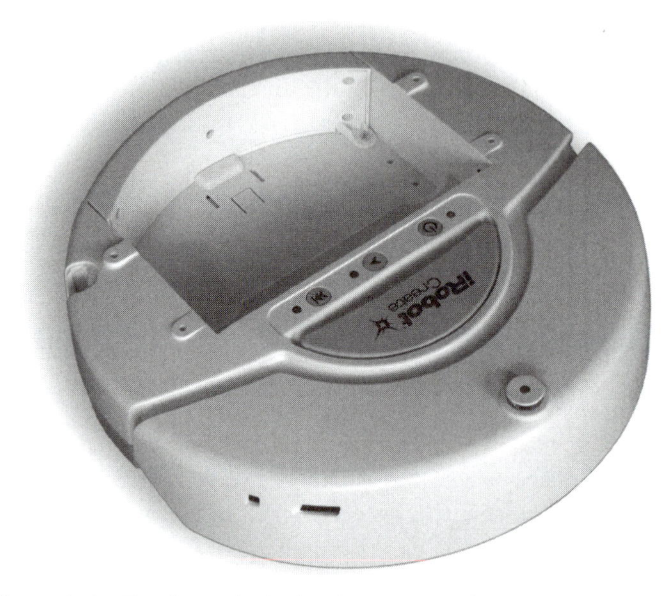

Figure 1-6 The Create by iRobot is a preassembled robot that is ideal for students and hobbyists.

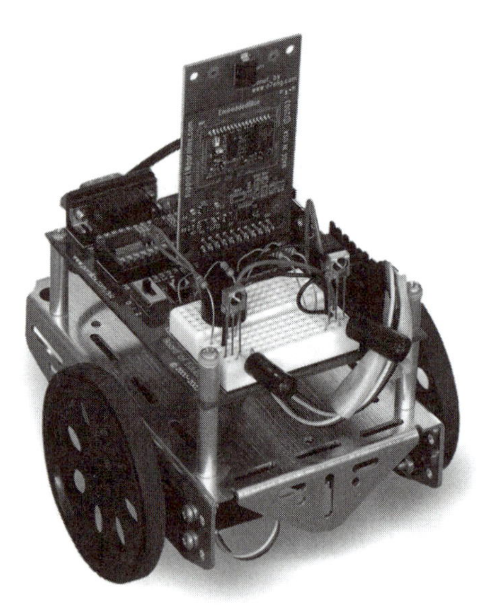

Figure 1-7 The Boe-Bot by Parallax requires little assembly and supports a Bluetooth module for remote access.

Figure 1-8 The LEGO Mindstorms NXT is a powerful and affordable robot that includes built-in Bluetooth capabilities.

The remaining robots supported by MSRS, which are listed in Table 1-1, range from small to large, affordable to expensive. Determining which robot to use will depend on your situation. If you are new to robotics, a robot such as the Create might be the ideal robot for you. If you have a little experience and are comfortable assembling small electronics, the Boe-Bot, LEGO Mindstorms NXT, or fischertechnik kits might be a better option. For serious roboticists, there is the Kondo KHR-1 or the Pioneer 3DX. There is also the robotic arm made by industrial robot maker KUKA. This six-axis, jointed-arm robot is featured in several tutorials available on the KUKA Web site (*http://www.kuka.com/usa/en/products/software/educational_framework/*).

Note The prices in Table 1-1 are subject to change; you should check the URLs in the Description column for the latest prices.

Table 1-1 **Supported Robots**

Robot	Description	Price
iRobot Create or Roomba	Several different Roomba models are available, but you might prefer to work with the Create robot (see Figure 1-7 earlier) because it was designed specifically for programmers interested in robotics. You can purchase the Create robot here: *http://store.irobot.com/family/index.jsp?categoryId=2591511*.	$129.99 and up
Parallax Boe-Bot BASIC Stamp Robot	The Boe-Bot is a wheeled robot offering sensors such as photo-resistors, bumpers, and infrared sensors. By using one of the add-in kits, you can turn your Boe-Bot into a tank or a crawler, or you can equip it with a camera and use it for surveillance. You can purchase the Boe-Bot serial or USB version from the Parallax Web site: *http://www.parallax.com/detail.asp?product_id=28132*.	$149.95

Table 1-1 Supported Robots

Robot	Description	Price
LEGO Mindstorms NXT	This is LEGO's latest robot kit. It features a 32-bit processor and allows you to control your robot remotely using Bluetooth. It also includes several sensors such as a sound sensor, ultrasonic visual sensor, and improved light and touch sensors. You can purchase the NXT from the LEGO Web site: *http://shop.lego.com/Product/?p=8527*.	$249.99
fischertechnik	This German-based company offers several computing robotics kits that allow you to build whatever kind of robot you prefer. To be compatible with MSRS, the kit must include the 16-bit ROBO interface. To order one of the robot kits, go here: *http://www.fischertechnik.com/html/computing-robot-kits.html*.	$332.99 and up
Kondo KHR-1	This Japanese humanoid robot is able to perform kung-fu fighting and acrobatics. Intended to be used as serious competition in robot wars, this robot is expensive and can be purchased here: *http://www.robotshop.ca/home/suppliers/kondo-en/index.html*.	$1,632.09 price can vary depending on reseller
MobileRobots Pioneer 3DX	This wheeled robot from MobileRobots Inc. allows you to execute MSRS code directly on the robot by using either the built-in 32-bit processor or an onboard laptop computer. It also includes several sensors such as bumpers, grippers, laser rangefinders, and a compass. You can get more information about this robot here: *http://www.activrobots.com/ROBOTS/p2dx.html*.	

One important thing to keep in mind is that MSRS is not just for hobbyists. It can be used to power some of the most sophisticated robots in the world. The robots listed in Table 1-1 represent only a small fraction of the robots available on the market today. They are just the ones that MSRS has built-in services for. There are other robots in which MSRS services have been written. For example, Robotics Connection, maker of the Traxster robotic kit (*http://www.roboticsconnection.com/pc-15-3-traxster-robot-kit.aspx*), offers MSRS services downloadable from its Web site. This rugged and fully expandable robot includes tank treads and can be a good starter kit for those new to robotics.

Gearing up for the Urban Challenge

Since 2004, DARPA has sponsored an annual autonomous vehicle race. The original race, which took place in the Mojave desert, was the longest distance competition for robots in the world. Unfortunately, no vehicles were able to cross the finish line that first year. But the next year, many teams returned with re-engineered designs, and this time there was a winner. The Stanford team completed the 132-mile rugged terrain course in a little less than seven hours. At this point, DARPA considered the challenge met and wanted to up the ante a bit. In 2006, DARPA announced a new competition—an urban-based one, in which the autonomous vehicles would have to deal with other challenges, such as traffic signals and lost GPS signals.

Princeton University sponsored teams for both the 2004 and 2005 Grand Challenge, and it has a team in place for the 2007 Urban Challenge. The team was made up of undergraduate students majoring in electrical engineering, chemical engineering, mechanical engineering, and computer science. This team was one of 35 that met the requirements to compete in a Qualification Event held in Victorville, California. Just to be eligible to compete in this event, the team had to pass a preliminary site visit by DARPA officials and then complete four required missions. Although, ultimately, they did not qualify for the final event, theirs has been quite an accomplishment nevertheless.

The team has chosen MSRS as the software to power their autonomous vehicle, thus making it the largest and most complex deployment of MSRS to date. The software system used to operate the vehicle will use 25 services that run on five distributed servers. The services must be able to handle not only navigation but must also avoid all other vehicles and obey traffic signals.

Depending on the robot you are working with, the time required to assemble the robot and configure it to work with MSRS will vary. Some of the robots supported by MSRS (such as the iRobot Create) come pre-assembled and involve very little configuration. Other robots (such as the LEGO Mindstorms NXT and Parallax Boe-Bot) can take more time to assemble but provide an opportunity to learn more about how the robotics hardware works.

Tip If you are new to the field of robotics and you are trying to decide which robot kit to purchase, consider this: The Create by iRobot will be the simplest to work with because no assembly is required. The LEGO Mindstorms NXT requires considerable assembly, but it also offers a wide range of possibilities for how the robot looks and behaves. The Boe-Bot by Parallax is the most complex robot to work with, but it offers valuable insight into how electronic-based robots work. This can be very valuable for readers who are interested in building their own robots from scratch one day.

What If You Want to Work with Unsupported Hardware?

MSRS comes with prebuilt services that allow you to work with the list of supported robots. However, you are able to build applications for unsupported hardware by writing your own services. The unsupported robot or device will need to expose a remote communications interface that is used to control the motors and read data from sensors. If your hardware does not already have this, you'll have to create an interface by using the coding platform that works with your robot or device. For readers interested in working with a unsupported hardware, refer to Chapter 7, "Creating a New Hardware Interface."

Challenges of Robotics Programming

.NET developers already familiar with Visual Studio and the .NET Framework will likely adjust to using MSRS the quickest. However, robotics programming is very different from traditional types of programming, especially Web development. Even though some developers may have worked somewhat with asynchronous processing, it's unlikely that they've had to manage as many asynchronous tasks as a robotics application might require.

Web development with the .NET Framework typically involves one or more concurrent users entering and retrieving information from a Web page. Even though the information is dynamic and typically fed from a relational database, the processing is generally sequential, and one process typically does not start until another one completes. With robotics programming, multiple sensors could be receiving data at any and all times. To accomplish autonomous behavior, in which the robot is able to roam about freely, the code needs to be able to handle reading sensor data and then directing the robot to do something. All of these actions must occur at the same time.

What If You Want to Execute Code on the Robot?

Rather than executing code on a PC, it's possible to execute MSRS programs directly on the robot. Unless you have a robot with an embedded 32-bit processor (such as MobileRobots's Pioneer 3DX) capable of running the .NET Framework, your robot or device should be able to run Windows CE 5.0 or higher. In this scenario, the development will still be done on a laptop or desktop machine. Once tested, the code can be moved to the CE device, thus not requiring a connection between the development PC and CE device. For more information about developing robotics applications targeting the .NET Compact Framework, refer to *http://msdn2.microsoft.com/en-us/library/bb483099.aspx*.

Summary

- In the past few years, there has been a noticeable increase in the need for professional and domestic robots. At the same time, it has been difficult for new programmers and engineers to enter the field of robotics. MSRS offers a standard platform that can be used by students as well as robot enthusiasts.
- The Web-based runtime offered by MSRS incorporates technologies used to power the World Wide Web with technologies used to manage asynchronous processes. Using these technologies, the runtime is able to offer a flexible, yet efficient, platform.
- MSRS offers helpful tools such as the application development tool VPL. This graphical-based tool lets you build applications by dragging and dropping activities onto a design surface. MSRS also includes a simulation engine and editor that allows you to build elaborate simulations using code that can then be used to operate actual robots.
- MSRS supports several different robots that range from simple to complex, affordable to expensive. The one you select depends on what goals you are trying to accomplish and how much money you are willing to spend.

Chapter 2

Understanding Services

Services are the basic building blocks for robotics applications built with Microsoft Robotics Studio (MSRS). This chapter will focus specifically on what makes up a service and how it is used to manage and control a robot's sensors and actuators. You will learn how services manage state and handle sending and receiving messages. You will also learn how services handle subscriptions and communication with other services. Finally, you will learn how a service handles output and embeds a resource such as a reference to an external dynamic-link library (DLL) file.

Defining a Robotics Application

Before defining a service, let's look at what makes up a robotics application. In terms of the robots you will be working with, the robot is just a physical device that is typically mobile and contains several sensors used to monitor the environment. Sensors are used to measure things such as the intensity of light or the distance from the robot to another object. Without any programming, the addition of a sensor is useless. You must have some piece of software that collects raw data from these sensors and then processes the results.

In addition to sensors, a robot also contains output devices known as actuators. A small speaker or light-emitting diode (LED) would qualify as an actuator. Actuators allow the robot to respond to the environment based on the information received from the sensors.

For example, a small, wheeled robot may have a bumper sensor, and you could create a program that read data from this sensor. While driving the robot, if the data returned from the sensor indicated that the bumper had been pushed, you could assume the robot had hit an object. You could add code to the program that stopped the robot from driving and then alerted someone by flashing one of the LED lights or sending a beeping sound to the speaker. In the example just described, the data from the bumper sensor was the input, and the signal to stop driving and flash a light or send a beeping sound was the output.

A robotics application is similar to other types of applications in that it receives input, performs some kind of processing, and then returns results as output. What makes the robotics application so unique is that this typically involves more than one form of input and output, and they can all be functioning at the same time. Multiple sensors and actuators can be located on a single robot. One sensor may be collecting data at the same time that output is sent to an actuator. All of these operations need to happen synchronously, or at the same time, in order for the robot to remain responsive at all times.

What Is a Service?

To accommodate the special needs of robotics applications, MSRS uses a service as the key object. Each Web-based service contains the code needed to perform one or more functions, such as reading data from a single sensor or sending an output signal to an actuator. The service can also be used to communicate with another service or external software. A robotics application consists of multiple services that work together to achieve a common task—operating the robot. For example, Figure 2-1 is an application diagram that represents the services used to operate a LEGO Mindstorms NXT.

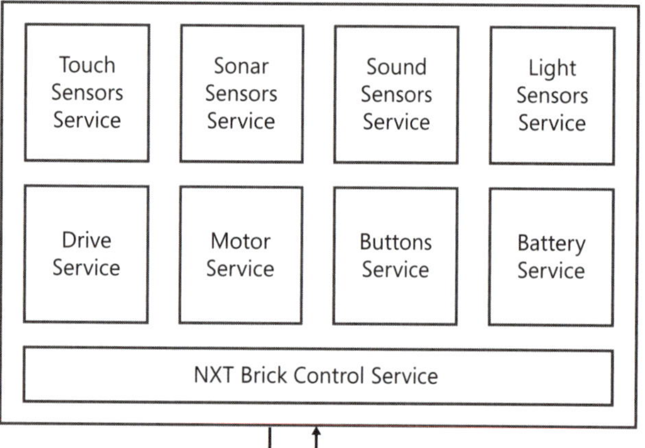

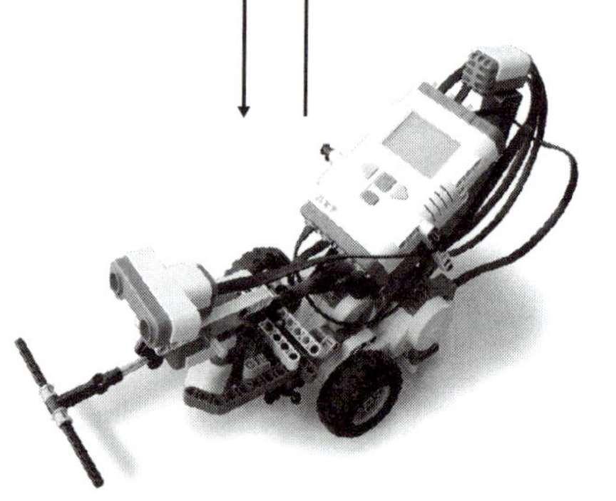

Figure 2-1 Multiple services work together to support the operation of a LEGO Mindstorms NXT.

Tip In addition to the material in this chapter, the reader should review the step-by-step service tutorials provided with MSRS. The latest version of the tutorials can be found online at *http://msdn2.microsoft.com/library/bb483065*.

Note This chapter will not repeat the steps found in the tutorials, but it will instead serve as a guide to understanding the concepts demonstrated through them. None of the tutorials requires robotics hardware.

Creating a New Service

MSRS offers several command-line tools (see Table 2-1) that can be used to interact with the Decentralized Software Services (DSS) runtime services. Prior to the MSRS 1.5 release, developers had to create a new service manually using the DssNewService command-line tool. MSRS now features a Visual Studio template (see Figure 2-2) that allows you to create a new service using Visual Studio 2005. (Though the procedures and programs in this book were tested in Visual Studio 2005, they will also work with Visual Studio 2008.)

Table 2-1 Command-Line Tools Used to Work with DSS

Name	Description
DssHost	Used to start a DSS node on a port and run one or more services.
DssNewService	Used to create a skeleton for a new service. This would include the Visual Studio solution and project files. This utility provides an alternative and programmatic way of creating projects. The other option is to use the Visual Studio template, included with MSRS, for creating new DSS services.
DssInfo	Used to get information about a service such as all the services and contracts associated with an assembly.
DssProxy	Used to generate proxies and transforms. A proxy is an assembly that exposes the contract information for a product. This utility is rarely called directly. Instead, it is used as part of the build process when creating a new service.
DssProjectMigration	Used to migrate a project from a previous release version. This was particularly important in the early release days of MSRS when a new release was generated as often as once a month.
DssDeploy	Used to package all the files that go with a service and create an executable that can be used for deployment.

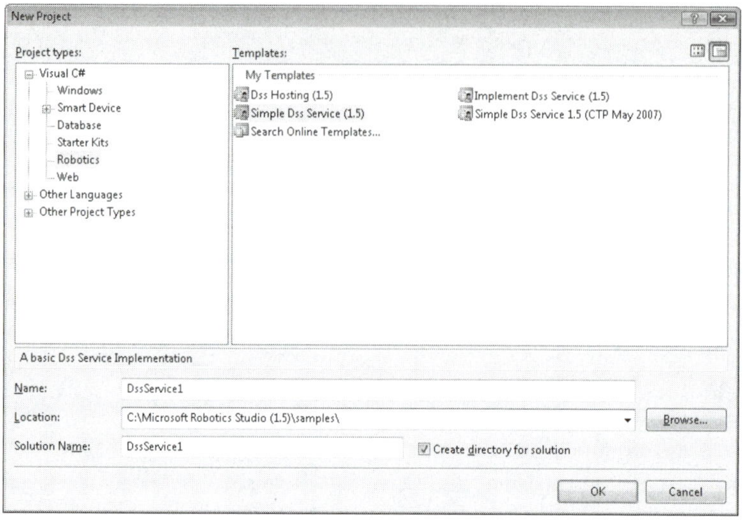

Figure 2-2 Included with MSRS is a Visual Studio template that creates the basic structure for a DSS service.

Starting a Service

Before we examine the code to create a new service, let's look at how services are started. The DssHost.exe command-line tool (listed in Table 2-1) is used to start a new DSS node on a port and run one or more services. In Chapter 1 you learned that DSS is one of the key technologies used by the MSRS runtime. A DSS node is the environment or context in which services will run, and it must be started on a port before any services can run on that node. The easiest way to start a DSS node is by clicking the Run DSS Node menu item under Microsoft Robotics Studio on the Windows Start menu.

Note It may take a few minutes for the DssNode executable to run on your development machine. Depending on your operating system and what antivirus tools you are running, the first time you run DssNode on your computer, you may be asked to grant access to MSRS. For users running Windows Vista, there is the additional constraint that you have to use a registered port. Refer to the following MSDN document online for additional security configuration information: *http://msdn2.microsoft.com/en-us/library/bb870571.aspx*.

Warning Do *not* close the command window that appears because this will cause the DSS node to stop, and you will not be able to access your services until you start the node again.

After executing DssNode, you will see a command window appear, and then your Internet browser will open and point to the following URL: *http://localhost:50000*.

The Internet browser should open to the System Services Web page for MSRS. From there you can click Control Panel on the left panel to see a list of services available on the machine (see Figure 2-3). To see a list of services that are running, click Service Directory on the left panel.

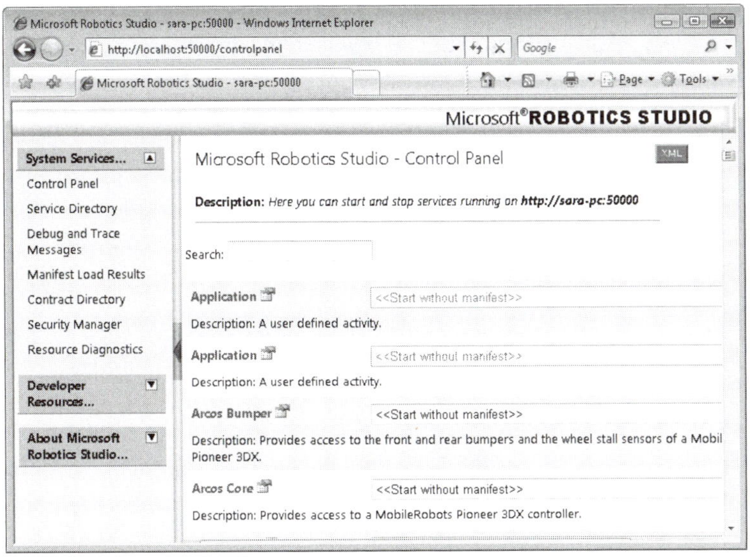

Figure 2-3 The MSRS Control Panel allows you to see a list of services available for the node.

Alternatively, you could start a DSS node manually to listen for HTTP requests on port 50000 and SOAP requests on port 50001 by going to a command prompt, navigating to the bin subdirectory where your MSRS files were installed, and typing the following command:

```
Dsshost -p:50000 -t:50001
```

> **Tip** The MSRS installation includes a Command Prompt link that opens a command prompt and then adds "bin" to the search PATH so you do not have to navigate to the bin folder yourself. From this command prompt, you can execute all the command-line tool utilities listed in Table 2-1.

This command will start a DSS node with no *manifest* or services loaded. Refer to Table 2-2 for a list of arguments that you can use with the DssHost executable.

> **Manifest** A manifest is an Extensible Markup Language (XML) file that includes a list of services to be started. This is typically used when starting a DSS node. The manifest is specified by passing the Uniform Resource Identifier (URI) or path as an argument for the DssHost executable.

Table 2-2 **Parameter Listing for the DssHost Executable**

Long Name	Short Name	Description
/manifest	/m	URI or path that points to the initial manifest(s) to load.
/dll	/d	URI or path that points to the initial DSS service DLL to load.
/contract	/c	URI or path that points to the initial DSS *contract* to load.
/port	/p	Integer value that contains the Transmission Control Protocol (TCP) port used to listen for Hypertext Transfer Protocol (HTTP) requests. 50000 is the port number typically used because this does not conflict with any known ports.
/tcpport	/t	Integer value that contains the TCP port used to listen for SOAP requests. 50001 is the port number typically used because this does not conflict with any known ports and is a different port than the one used for HTTP requests.
/hostname	/h	String value that contains the host name that identifies the host.
/security	/s	URI or path that points to a security settings file.
/verbosity	/v	Verbosity level. Can be Off|Error|Warning|Info|Verbose. *Info* is the default value.
@file	@file	Name of a text file that contains a list of arguments. This is a good alternative to repeatedly typing in the same arguments.

Contract A contract contains information that other DSS services need in order to use the service. The contract is associated with an HTML file, but you cannot access this file directly. Instead, you can use the DssInfo command-line tool to view the contents of a contract. For more information about the contract, refer to the following URL: *http://msdn2.microsoft.com/library/bb648746*.

The MSRS Control Panel contains a list of all services available on the node. Even if you have never created a service, you will see several listed. Some of these are services used to run MSRS, some are used to run the supported hardware platforms, and some are generic services that can be used to support any hardware platform. Take the time to browse through this list of services and become familiar with the built-in services that MSRS provides.

Creating a Service Project

To create a new service, open Visual Studio 2005, choose File, New, and then click Project. In the New Project dialog box (see Figure 2-2), expand the node for the language you prefer, select the Robotics node, and then select the Simple DSS Service (1.5) project template. In the Name text box, type a name for the new service (or use the default service name of DssService1 to make it easier to follow our examples), enter a location for the code files, and then click OK. This will create the solution and files you need to build a simple DSS service. These files include class files and an .xml manifest file.

Assuming you used the default name for the service, your class files will be named DssService1.cs and DssService1Types.cs. DssService1.cs is the implementation class, and this is where you will place the code that reads data from sensors and sends commands to your robot. DssService1-Types.cs is the contract class, and this is where you will return information about the service such as the *state*. The *contract* class will also handle any requests to drop or create the service.

> **State** You can think of services as active documents that contain useful information. At any point, you can open these documents and see what information they contain. When you do this, you access the service's state, which is a representation of the service at the time it is requested. For example, if you have a service that returns data from a sensor, then the state for that service would be the data read from the sensor at the time you requested the state.

Without adding any code, you can build the DssService project created with the Visual Studio template. By default, the state for a service is empty, which means that no information is returned when someone accesses that service. So far, the DssService project does not communicate with any devices, and it will not return any state.

If you wanted the DssService project to do something simple, then you could have it return the phrase "Hello World." In this case, the phrase "Hello World" is the service's state, and this phrase is returned when someone accesses the service. To accomplish this, you would need to add the following code to the DssService1Types.cs file (inside the public class definition for the *DssService1State* class):

```
private string _outputmsg = "Hello World";
[DataMember]
public string OutputMsg
{
    get { return _outputmsg; }
    set { _outputmsg = value; }
}
```

The *DataMember* attribute indicates that the public properties that follow should be serialized as XML. Serialization is the process of representing a .NET object as a stream of bytes, such as a SOAP message. Alternatively, instantiating a .NET object from the data in a SOAP message is known as *deserialization*.

The class definition for *DssService1State* is marked with the *DataContract* attribute. This indicates that the objects within the *DssService1State* class should be included in the DSS proxy dll that will be generated for this service.

> **Tip** By default, the Visual Studio template available with MSRS will return the XML as a SOAP message. If you want to return the XML as HTTP and avoid the overhead of a SOAP message, you can add HttpGet to the PortSet. This will entail adding additional code to your service. For more information about how to do this, refer to the Service 1 tutorial available at *http://msdn2.microsoft.com/library/bb905437*.

To understand this better, let's take a look at the SOAP response message for the "Hello World" DssService project. To see this for yourself, you will need to successfully build and run the DssService1 project by pressing F5. This will build the project and create an assembly in the MSRS bin directory. By default, the project properties for the Visual Studio template will also execute DssProxy to build a *DSS proxy assembly*. Finally, the debug properties for the project will execute DssHost to start the node and load the DSS service.

> **DSS proxy** A DSS proxy is an assembly used to represent a service. Other services will use this proxy to communicate with this service rather than communicate with the service directly. The proxy is generated when the service is compiled, and it contains stubs for all public interfaces.

After pressing F5 to build and start the service, leave the command window open, and use your Internet browser to go to the URL for this service (the URL should be *http://localhost:50000/DssService1*). The SOAP message that appears should look like the following:

```
<s:Envelope xmlns:s=http://www.w3.org/2003/05/soap-envelope
   xmlns:wsa=http://schemas.xmlsoap.org/ws/2004/08/addressing
   xmlns:d="http://schemas.microsoft.com/xw/2004/10/dssp.html">
<s:Header>
  <wsa:To>http://127.0.0.1:1153/</wsa:To>
  <wsa:Action>http://schemas.microsoft.com/xw/2004/10/dssp.html:GetResponse
  </wsa:Action>
  <d:Timestamp>
 <d:Value>2007-07-10T13:16:28.0938784-05:00</d:Value>
  </d:Timestamp>
  <wsa:RelatesTo>uuid:99cf78b1-dabf-4b7d-8c6b-14124cbfc023</wsa:RelatesTo>
  </s:Header>
<s:Body>
<DssService1State xmlns="http://schemas.tempuri.org/2007/07/dssservice1.html">
    <OutputMsg>Hello World</OutputMsg>
</DssService1State>
</s:Body>
</s:Envelope>
```

Note that the body of the SOAP message includes an element named DssService1State. The XML namespace associated with this element is the contract name for the service. The namespace is simply a unique string used to identify the contract. It happens to look like a valid URL, but it is not required to be a path to an actual file.

> **Tip** When creating your own robotics applications, especially if you plan on distributing them commercially, you need to change the prefix of the namespace to something other than schema.tempuri.org. This will ensure that the contract for your service is unique and does not conflict with any other services. To change the namespace, change the Identifier variable located in the contract class. (This is found in the DssService1Types.cs file.) The namespace for the contract should also be changed in the DssService1.manifest.xml file.

Sending and Receiving Messages

In the last section you learned how to create a new service that simply had the state consisting of a string with the value "Hello World." To accomplish this, the phrase was posted to a *DSS port*. Messages, which represent a DSS operation on a port, allow other services to read and update the state of a service. In the last example, the phrase "Hello World" represented the state of the service. This phrase was sent to a port, and the result was a SOAP message that appeared in an Internet browser.

> **Port** A DSS port provides a way for services to communicate by passing messages back and forth. A service can specify one or more ports, and each port contains a data structure that holds the messages sent to the port. By default, each service has access to one main port. It is possible to define an internal port that can be used for updating a privately held state.

Anatomy of a Message

The basic function of every service is to send and receive messages. This is what enables services to communicate with each other. Messages are strongly typed, and message types are specified within a service. DSS requires that a message must contain the following three elements:

- **Action** Typically, this is a verb that informs the runtime what operation should be performed. For example, the verb *Get* indicates that the service should get the latest version of the state, and *Update* indicates that the service should update the latest version of the state.
- **Body** This is the body of the request. It does not represent the text to be sent (i.e. "Hello World" because that is the state). The body represents the XML found in the Microsoft.ServiceModel.Dssp.GetRequestType instance. For example, the following XML is the body for a GetRequest:

  ```
  [DataContract]
  [XmlType(Namespace = "http://schemas.microsoft.com/xw/2004/10/dssp.html")]
  [XmlRoot("GetRequest", Namespace =
     "http://schemas.microsoft.com/xw/2004/10/dssp.html", IsNullable = false)]
  ```

- **Response Port** This is the port to which the response should be sent. If specified, a response should be sent regardless of whether the action results in a success or failure.

The "Hello World" phrase is returned by something known as a DSS *transform* file. The transform file, along with the service proxy, is generated when the service assembly is compiled. The state items for a service will pass through calls to the transform before being returned to the service.

> **Transform** A transform is an assembly that is created when the DSS proxy is generated. The proxy exposes the service contract, and the transform provides a mapping between the service and the proxy. It acts as a bridge between the service and proxy assemblies.

The proxy and transform files are generated automatically when you use the Visual Studio template. To see where this is called, on the Project menu, choose *project_name* Properties. Click the Build Events tab and you see the command line for the DssProxy utility in the post-build event command line. By default, the assemblies generated are stored in the bin subdirectory for the MSRS installation.

> **Tip** If your service is running on the DssHost, you will not be able to compile your service until the DssHost process has been stopped. You can do this by simply closing the command window that is launched when you run the DSS node.

When compiling a service, MSRS will generate three assemblies: one for the service, one for the proxy, and one for the transform. They will be named using the following naming convention: <*service name*>.Y<*year*>.M<*month*>.<*Proxy, Transform, or blank*>.dll. For example, the following three files will be generated based on the date the service named DssService1 is compiled:

- C:\Microsoft Robotics Studio (1.5)\bin\DssService1.Y2007.M07.dll
- C:\Microsoft Robotics Studio (1.5)\bin\DssService1.Y2007.M07.Proxy.dll
- C:\Microsoft Robotics Studio (1.5)\bin\DssService1.Y2007.M07.Transform.dll

Using a PortSet

To understand how MSRS handles sending and receiving messages through DSS ports, let's take a closer look at the code that is generated when you create a new service using the Visual Studio template. In the DssService1.cs file, there is a declaration for the main port (as shown below):

```
[ServicePort("/dssservice1", AllowMultipleInstances=false)]
private DssService1Operations _mainPort = new DssService1Operations();
```

This code indicates that the main port does not allow multiple instances and that the service will be located in a subdirectory named dssservice1. Given this information, you can determine the URL of the service, *http://localhost:50000/dssservice1*. If the port allowed for multiple instances, each instance would be appended with a globally unique identifier (GUID).

The code also indicates that the port can perform Decentralized Software Services Protocol (DSSP) operations defined in the *DssService1Operations* class. You might recall from Chapter 1 that DSSP is the SOAP-based protocol used by DSS to handle messages. The *DssService1-Operations* class is located in the DssService1Types.cs file and looks like the following:

```
[ServicePort()]
public class DssService1Operations : PortSet<DsspDefaultLookup,DsspDefaultDrop, Get>
{
}
```

The *DssService1Operations* class defines a PortSet, and this PortSet is used to indicate what types of DSSP or HTTP messages are allowed. All services are required to include the *Dssp-DefaultLoopkup* operation, which defines a default message handler for a Lookup message. The Lookup message is used to return the service context, which contains information about how to communicate with the service. *DsspDefaultDrop* is not required, but it allows the service to support the *Drop* message. Finally, the *Get* message allows the service to respond with its current state.

Get is just one of several DSSP operations allowed (refer to Table 2-3 for a list of additional DSSP operations). It will return state in the format of a SOAP message as required by DSSP. To return the state as raw XML, you would need to add an HTTP operation such as *HttpGet* to the PortSet list. An example of this is provided in Service Tutorial 1 and is available at *http://msdn2.microsoft.com/library/bb905437*.

Table 2-3 DSSP Operations Provided with MSRS

Operation	Default Implementation	Description
Create	DsspDefaultCreate	Creates a new service. You do not have to call this directly because it is handled by the *DsspServiceCreationPort*, which is specified in the default service constructor.
Delete	DsspDefaultDelete	Deletes the part of the state identified in the *Delete* operation. This only deletes state and not the service itself.
Drop	DsspDefaultDrop	Shuts down the service. This must be the final message sent to the service.
Get	DsspDefaultGet	Used to get the entire state for a service. If the state consists of multiple elements, then they will all be returned. To access specific elements, you need to use the *Query* operation.
Insert	DsspDefaultInsert	State that is included with the request is added to the state belonging to the service.
Lookup	DsspDefaultLookup	Returns the service context for a service. This operation is required for all messages.
Query	No default provided	Retrieves state based on a specific parameter-based request. Only a specific portion of the services state is returned. DSSP does not require a structured query language to be used. The service performing the query must know what the service containing the state expects in terms of schema and query language.

Table 2-3 **DSSP Operations Provided with MSRS**

Operation	Default Implementation	Description
Replace	DsspDefaultReplace	Replaces all elements in the service state.
Submit	No default provided	Similar to an execute statement, submit will perform computations that do not alter the state of a service.
Subscribe	No default provided	Allows a service to receive event notifications regarding state changes with another service.
Update	DsspDefaultUpdate	Used to specify a portion of the state to update. The update request will perform a delete and insert wrapped in a transaction to ensure both operations succeed.
Upsert	No default provided	Combination of an Insert and an Update. If the state already exists, then the state is updated. Otherwise, the state is inserted.

Table 2-3 lists the DSSP operations that can be added to a PortSet. The DSS Service Model provides built-in classes that define and implement default code for some of these operations. You can get started by using these defaults instead of adding the code to your service. Ultimately, you will only want to use the defaults when the state for a service is not known or when the implementation is generic. In most cases, you will only use the defaults for the *Lookup* and *Drop* operations.

Keep in mind that you can add your own custom messages, which derive from the *DsspOperation* class. Most services you write will include at least one custom operation implementation. An example of this is seen with the *IncrementTick* message defined in Service Tutorial 2 (which is available at *http://msdn2.microsoft.com/library/ bb905437*). In addition to the DSSP operations, you can also use HTTP-based operations (as shown in Table 2-4).

> **More Info** More information about interacting with services using SOAP and HTTP is available through the following URL: *http://msdn2.microsoft.com/library/bb727264*.This includes how to access services using a Web browser.

Table 2-4 **HTTP Operations Provided with MSRS**

Operation	Changes State? (Yes/No)	Description
HttpDelete	Yes	Used to delete the state element requested.
HttpGet	No	Returns state using a URL. This works the same as when you retrieve documents by browsing the Web.
HttpOptions	No	Represents a request for information about the communication options available.
HttpPost	Yes	A post passes parameters inside the actual HTTP request message. This is typically used when dealing with HTML form data.

Table 2-4 HTTP Operations Provided with MSRS

Operation	Changes State? (Yes/No)	Description
HttpPut	Yes	A request that the state of the service is replaced with the data included in the request.
HttpQuery	No	Similar to an *HttpGet*, but with query parameters passed in with the URI.

Using Service Handlers

A service handler is responsible for handling incoming messages on a port. A service can have more than one service handler, and each one will pertain to a different DSSP, HTTP, or custom operation. The only exceptions to this are the *DsspDefaultLookup* and *DsspDefaultDrop* operations because these are typically handled by the DSS runtime but can be overwritten by the service if needed.

The Visual Studio template, which is used to create a new DSS service, provides only one service handler. The service handler (shown as follows) is used to handle DSSP *Get* messages:

```
[ServiceHandler(ServiceHandlerBehavior.Concurrent)]
public virtual IEnumerator<ITask> GetHandler(Get get)
{
    get.ResponsePort.Post(_state);
    yield break;
}
```

The *GetHandler* does not do much. It simply posts the state as a SOAP message to the response port. The code in this handler could contain additional code that performs computations or sends messages to other services.

Each service handler allows you to define how it will be processed. For example, *GetHandler* is set to run concurrently. This behavior should be set for handlers that do *not* modify state. For handlers that modify state, the *Exclusive* service handler behavior attribute should be used. Other attributes include *Teardown*, which is executed once and shuts down the service, and *DsspOperationDefault*, which picks the behavior based on the operation type.

Managing State

In the preceding section, titled "Sending and Receiving Messages," you saw the basic code requirements for handling messages. Messages are used for requests or responses dealing with state. Sometimes, it may be necessary to change the state associated with a service. In this section, we will look at the ways you can manage state.

Posting Data to an Internal Port

In the last section, the *Get* operation was used to retrieve the state for a service. The state was posted to the main port for the service, which resulted in a SOAP message appearing in the Internet browser. Alternatively, state can be posted to an internal port. This is useful in cases in which data needs to be collected and saved until it is requested. Such data can be stored in a privately held version of the state. The state will only be maintained as long as the service is running.

To understand how this might work, consider a continually operating service that is responsible for monitoring a network port for all incoming HTTP requests. When a request is received, it could temporarily store information about the request on an internal port. The state data would not be posted to the main port until it is requested by a user or another service.

Another example is a long-running service that is responsible for reading a serial port or parallel port and storing the data returned in a privately held state. The data would only be posted to the main port when it was requested.

To post data to an internal port, you need to declare state variables to store the internal data. You also need to declare an additional port, such as in the following code where a port named MyInternalPort is created:

```
Port<string> MyInternalPort = new Port<string>();
```

You then need to create a new message that is derived from the *Update* generic class. You will also create a new operation type and service handler for the operation type you are creating. This new operation type will be added to the list of operations supported for the main PortSet. For example, a service that supported an operation named *GetURL* would have a main operations port that looked like the following:

```
[ServicePort()]
public class MonitorRequestsOperations : PortSet<DsspDefaultLookup,
        DsspDefaultDrop, Get, GetURL>
{
}
```

You will also need to add an exclusive message handler for the new message. This is where you would add code to poll the HTTP port for requests. To post data to an internal port, you will need to utilize the Concurrency and Coordination Runtime (CCR) *Arbiter* class. The CCR is necessary to avoid conflicts when handling inbound requests. The *Arbiter.Receive* method is used to create a single item receiver. For CCR to manage the receiver task, it needs to be wrapped inside a call to the *DsspServiceBase.Activate()* method. For example, the following code can be used to post a string to the port named MyInternalPort:

```
MyInternalPort.Post(message);
Activate(Arbiter.Receive(true, MyInternalPort, MyInternalPortHandler));
```

Saving State to a File

In some cases, it may be beneficial to save the state of a service to a file so it can be retrieved later. This can be useful when you want to save settings to a configuration file. State can be persisted with the use of an Initial State *partner*. A partner is declared by using a *Partner* attribute. Services can have more than one partner or no partners declared.

> **Partner** A partner represents a service that is tied to another service. Partnerships are established to inform the MSRS runtime about relationships and possible dependencies between services. Some partnerships, such as ConstructorService, are generated automatically when you compile your service.

You can see whether a service is associated with a partner by selecting the Service Directory from the System Services Web page (see Figure 2-4). Each partner associated with a service will be listed in the Partners's column for the service instance. For example, the service named ServiceTutorial3 is associated with the following three services:

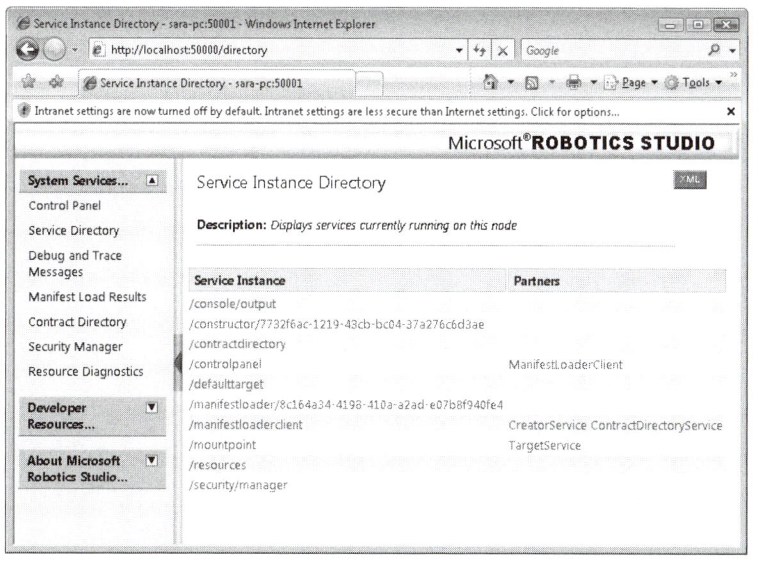

Figure 2-4 Service Instance Directory lists all services currently running on the node. From here you can see a list of partners associated with the service.

- **ConstructorService** This is used to initiate a new service for services that use the Create operation. A new instance of the ConstructorService will be created automatically for each service using a GUID (this is created at compile time).
- **PartnerListService** The Partner List Manager service is used to return a list of valid partners. This partner will be listed for all services regardless of whether a partner is included.

- **StateService** This service is used when there is a need to manage state. It is not necessary to include this service in the list of partners if state is only returned and not saved.

DSS allows the use of a specific kind of partner known as the Initial State partner. This type of partner is used to set the initial value of the state based on values retrieved from a state document. For example, the following code uses the *InitialStatePartner* attribute to declare a partner that will save the state to a file named MyState.Config.xml:

```
[InitialStatePartner(Optional = true, ServiceUri = "MyState.Config.xml")]
private MyServiceState _state = new MyServiceState()
```

Because the partner was declared with the optional tag, the service will start even if the MyState.Config.xml does not exist. The state document should reside in the store directory beneath the MSRS installation folder.

If the state does not already exist, then you need to add code to the *Start()* method. This method will execute every time the service is started. If no state document exists, then the code should set initial values for all state variables. For example, the following *Start()* method could be used to initialize the state for the DssService1 service created earlier:

```
protected override void Start()
{
 //Check to see if the state already exists.
 //If not, then we will initialize it
 if (_state == null)
 {
      _state = new DssService1State();
      _state.Message = "";
 }

 //Start the service
 base.Start();

}
```

In the example above, we were dealing with one state variable because only one variable was marked with the *DataMember* attribute. The state for a service can consist of multiple variables of differing types.

To save the state, you will need to add code that saves the state to the MyService1.config.xml file. The *SaveState()* method is part of the DSS service model, and it is available to your service application within the following .NET namespace:

```
using Microsoft.Dss.ServiceModel.DsspServiceBase;
```

You can reference methods in the DsspServiceBase by using the base reference, which represents the base class for service implementations. Code that saves the state should be added to an area of code in which the state changes value. You will then invoke the *SaveState()* method through the base reference, such as the following:

```
base.SaveState(_state);
```

Saving state can be useful when there is a need to maintain state data in between calls to a service. To see a step-by-step example of saving state to a file, refer to Service Tutorial 3, which is available at *http://msdn2.microsoft.com/library/bb483063*.

Communicating with Other Services

Through the use of partnerships, a service can call a method from another service and return the state from that service. One service (e.g. DssService2) can use the *Partner* attribute to establish a relationship with another service (e.g. DssService1). In order to do so, a reference needs to be added to the DssService2 project.

When adding a reference to another service, you need to browse to the location of the proxy assembly and not the actual service assembly (see Figure 2-5). You may then add a .NET namespace alias to the DssService2 implementation class. For example, the following code is used to define an alias for DssService1:

```
using svc1 = Robotics.DssService1.Proxy;
```

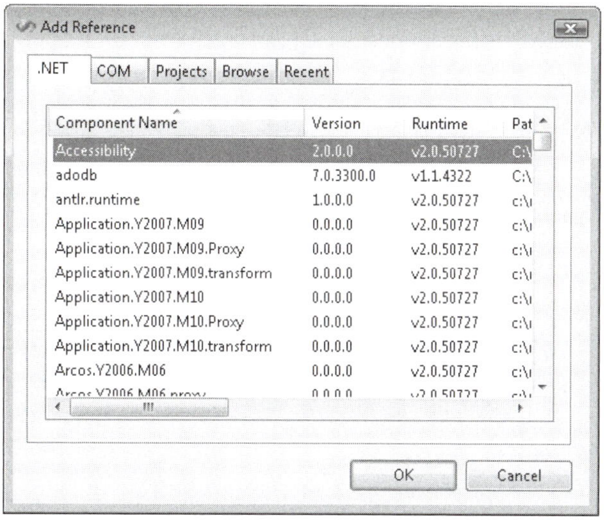

Figure 2-5 The Add Reference dialog box is used to add a reference to a partner service.

Once the reference is added, you can add code to establish the partnership. This is done by using the *Partner* attribute, and it will reference the alias declared. For example, the following code will declare a partnership with DssService1 using the namespace alias defined above:

```
[Partner("Svc1", Contract = svc1.Contract.Identifier, CreationPolicy =
        PartnerCreationPolicy.UseExistingOrCreate)]
```

```
private svc1. DssService1Operations _svc1Port = new
        svc1. DssService1Operations();
```

When communicating with other services, you need to be aware of how the two services interact with respect to threading. Otherwise, you could have one long running task that blocks the execution of a service. The CCR can help manage asynchronous tasks by taking advantage of the iterators available in C# 2.0. Using iterators, code that needs to execute asynchronously can still be written in a sequential fashion.

To maintain asynchronous behavior, you will need to use the *yield return* keyword, which is a new feature in C# 2.0. *Yield return* allows you to iterate through objects returned by a method call.

Additionally, the call could be wrapped inside the *Arbiter.Choice* method. For example, the following code uses the *Choice()* method to wait for the *updateRequest* to return with a success or failure:

```
yield return Arbiter.Choice(
               updateRequest,
               delegate(string response) { result = response; },
               delegate(Exception ex) { Console.WriteLine(ex); }
);
```

The *Arbiter* class (see Table 2-5), which is part of the CCR, provides several helper methods that can be used to coordinate requests and responses.

Table 2-5 Helper Methods Available in the *Arbiter* Class

Method Name	Description
Activate	Used to invoke a single item receiver that will excute every time an item is posted on the port
Choice	Allows the iterator to handle multiple outcomes such as success or failure
ExecuteToCompletion	Used to activate a task and then yield return to the iterator
Interleave	Can be used to protect a resource from concurrent access
JoinedReceive	Used to activate a multiple port receiever that can execute tasks in any order
JoinedReceiveWithIterator	Also used to activate a multiple port receiever, but also accepts an iterator handler
MultipleItemReceiver	Used to collect items from one or more ports and in the same PortSet and return the results in type-based collections
MultiplePortReceiver	Used to wait for a specific number of messages across multiple ports
Receive	Used to execute a single item receiver
ReceiveFromPortSet	Creates a single item receiver for a non-generic PortSet
ReceiveWithIterator	Creates a single item receiver for an iterator user handle
ReceiveWithIteratorFrom-PortSet	Also creates a singe item receiever for a iterator user handle when connecting to a non-generic PortSet

Another way to get asynchronous protection from the interleave is to use the *Arbiter.ExecuteToCompletion()* method. This helper method can be combined with the *SpawnIterator()* method, which is used to invoke an iterator based on a message handler while also passing in arguments.

Handling Subscriptions

A robot needs to receive data from its sensors on a continuous basis. For example, a robot needs to be informed instantly if one of the bumper sensors has hit an object. To accommodate this need, DSSP allows services to receive event notifications from other services. A service that reads data from a bumper sensor will need to notify another service when the state of the bumper sensor has changed. The ability to handle subscriptions is a critical component of MSRS. Without this feature, it would be impossible to operate and monitor robotics applications in real time.

With a subscription involving two services (e.g., Service A and Service B). Service A could provide the state and serve as the publisher. Service B, the subscriber, requests to be notified whenever the state for service A changes. You will need to add subscription code to both services. Service A will need to support subscriptions, and Service B will need to initiate the subscribe request. Service B will then receive an event notification when the state for Service A changes (see Figure 2-6).

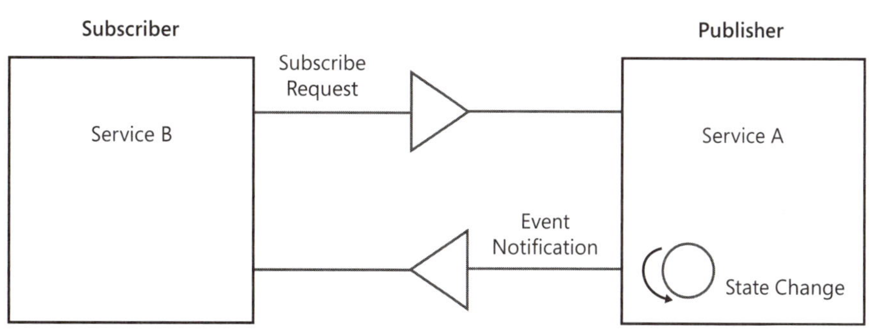

Figure 2-6 The relationship between a subscriber and publisher service.

Configure the Publisher

MSRS provides a system service named SubscriptionManager. You can see this service listed in Control Panel. The SubscriptionManager service is responsible for forwarding notifications to the appropriate subscribers and maintaining a list of those subscribers. To make it convenient to call the *SubscriptionManager* class from the publisher, you can define a namespace alias as shown in the following code:

```
using submgr = Microsoft.Dss.Services.SubscriptionManager;
```

The SubscriptionManager service functions as a partner service for each service that needs to support subscriptions. The SubscriptionManager service is declared using the *Partner* attribute. For example, the following code is used to declare the SubscriptionManager as a partner for the publisher service:

```
[Partner("SubMgr", Contract = submgr.Contract.Identifier,
        CreationPolicy = PartnerCreationPolicy.CreateAlways)]
private submgr.SubscriptionManagerPort _submgrPort = new
        submgr.SubscriptionManagerPort();
```

For a service to allow subscriptions, it must provide a subscribe message that supports SubscribeRequestType and SubscribeResponseType, such as the following:

```
public class Subscribe : Subscribe<SubscribeRequestType,
            PortSet<SubscribeResponseType, Fault>>
{
}
```

Once the message has been created, it must be added to the PortSet for the main operations port. There will also need to be a service handler for the subscribe operation. This handler is responsible for adding subscribers to the list using the SubscriptionService. Assuming that your handler accepts a parameter named sub, the *DsspServiceBase.SubscribeHelper* method can be placed in the SubscribeHandler:

```
SubscribeHelper(_submgrPort, sub.Body, sub.ResponsePort);
```

The last thing the publisher needs to do is send notifications to subscribers whenever the state has changed. Code to initiate a send notification must be added anywhere the state is updated. For example, the following code could be added to a ReplaceHandler:

```
base.SendNotification(_submgrPort, replace);
```

As you are adding code to send notifications, keep in mind that not all operations will generate an event. For example, *Get* and *Query* do not generate events. The operations that do generate events are as follows:

- Delete
- Drop
- Insert
- Replace
- Update
- Upsert

Configure the Subscriber

After the publisher service is configured, you will need to add code to the subscriber service. The first thing to do is reference the proxy assembly file for the publisher service. This is because services communicate with each other through their proxy files. The proxy assembly is referenced by right-clicking References from Solution Explorer and selecting Add Reference. You will then browse to the bin subdirectory for the MSRS installation and locate the proxy.dll file for the publisher service.

After the reference is added, you will need to change certain properties for the assembly to prevent it from being copied to the output directory for the subscriber service. By doing this, you will ensure that you are always referencing the correct version of the assembly. To change the properties, locate the assembly in the References folder, right-click the assembly, and then select Properties. Change the Copy Local and Specific Version properties to false.

To access the publisher class easily, you may add a namespace alias declaration to the subscriber's implementation class. You will need to add a Partner declaration, similar to the one created in the subscriber service. This will allow the subscriber service to access the *Subscribe* operation that was added to the publisher service. This is what allows the subscriber to receive notifications when the state for the publisher service changes. The last thing to do is to add code to the subscriber service that calls the Subscribe operation. To see a step-by-step example of configuring the publisher and subscriber, refer to Service Tutorials 4 and 5, available at *http://msdn2.microsoft.com/library/bb905438*. You can also take a look at the Atom/RSS Syndication service tutorial, available at *http://msdn2.microsoft.com/library/bb648747*.

Handling Output

MSRS allows you to format service state using an Extensible Stylesheet Language Transformations (XSLT) file. This gives the service developer control over how the state data is displayed. An XSLT file is used to transform an XML document or, in this case, the XML returned from a service. The result is a formatted HTML, XML, or text document. For more information about XSLT specification, refer to the following URL: *http://www.w3.org/TR/xslt*.

To transform XML, you will need to create an XSLT file. You can do this using Visual Studio by right-clicking the project in Solution Explorer and clicking Add and New Item. To format as HTML, the XSLT file will need to contain an output element that includes a method attribute set with a value of "html." For example, the following XSLT file can be used to format the single state item from a service named DssService1:

```
<?xml version="1.0" encoding="UTF-8" ?>
<xsl:stylesheet version="1.0"
     xmlns:xsl="http://www.w3.org/1999/XSL/Transform"
     xmlns:svc1="http://schemas.tempuri.org/2007/07/dssservice1.html">

  <xsl:output method="html"/>
```

```
  <xsl:template match="/svc1:DssService1State">
    <html>
      <head>
        <title>DSS Service 1</title>
        <link rel="stylesheet" type="text/css"
      href="/resources/dss/Microsoft.Dss.Runtime.Home.Styles.Common.css" />
      </head>
      <body style="margin:10px">
        <h1>DSS Service 1 - XSLT Example</h1>
        <table border="1">
          <tr class="odd">
            <th colspan="2">Service State formatted with XSLT</th>
          </tr>
          <tr class="even">
            <th>Output Message:</th>
            <td>
              <xsl:value-of select="svc1:OutputMsg"/>
            </td>
          </tr>
         </table>
      </body>
    </html>
  </xsl:template>

</xsl:stylesheet>
```

The stylesheet requires a reference to the service contract (*http://schemas.tempuri.org/2007/07/dssservice1.html*). If you do not know the URI for the contract, you can open the manifest.xml file for the service and retrieve it from the *dssp:Contract* element. This reference is bound to an alias, svc1, that is used throughout the stylesheet.

When the stylesheet is complete, you will need to add it as an embedded resource of the service assembly. This is done by right-clicking the XSLT file in Solutions Explorer and changing the Build Action to Embedded Resource. You will also need to add code such as the following:

```
[EmbeddedResource("Robotics.DssService1.TransformOutput.xslt")]
string _transform = null;
```

At this point you can reference it from the HttpGet message handler. You have to use the *HttpGet* operation because this returns XML and not a SOAP message. Code such as the following can be used to reference the newly embedded resource: *httpGet.ResponsePort.Post(new HttpResponseType(System.Net.HttpStatusCode.OK, _state, _transform));*

After compiling the service, the output displayed in an Internet browser should appear as formatted HTML (see Figure 2-7).

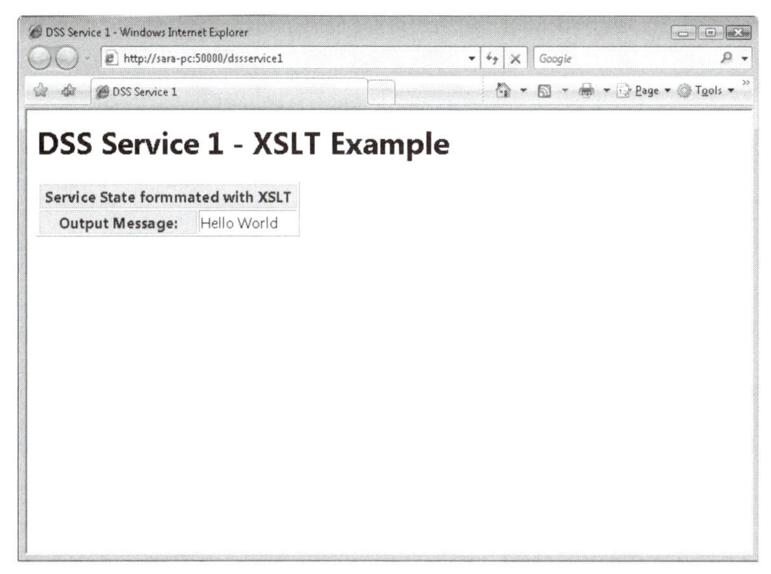

Figure 2-7 Screenshot of the state from DssService1 formatted with an XSLT file.

Handling Faults

If a sender requests a response, then one will be sent regardless of whether the operation was successful or not. Therefore, whenever an operation includes a response port, you should include the World Wide Web Consortium (W3C) SOAP fault body with the port set for that message. For example, the fault body is included in this Get operation, which was included in Service Tutorial 2:

```
public Get(Microsoft.Dss.ServiceModel.Dssp.GetRequestType body,
         Microsoft.Ccr.Core.PortSet<ServiceTutorial2State,W3C.Soap.Fault>
         responsePort) : base(body, responsePort)
{
}
```

DSSP provides a set of fault codes (see Table 2-6 for a list of DSSP fault codes) that can be used to check for specific conditions. For example, when performing an insert, you could first check to see if the state item already exists using the DuplicateEntry fault code. The following code provides an example of this scenario:

```
if ( /* the item already exists */)
{
 insert.ResponsePort.Post(Fault.FromCodeSubcodeReason(
FaultCodes.Sender, DsspFaultCodes.DuplicateEntry));
  yield break;
}
```

Table 2-6 DSSP Fault Codes Provided with the DSSP Service Model

Fault Code	Description
ActionNotSupported	The operation is not supported by the service.
DuplicateEntry	A duplicate entry already exists for the state item.
InsufficientResources	The operation failed due to insufficient resources.
MessageNotSupported	The DSSP body type supplied is not supported.
OperationCancelled	The operation was cancelled.
OperationFailed	The operation failed.
QueuingLimitExceeded	The operation failed because the queuing limit on the target port exceeded the limit.
ResponseTimeout	The response timed out.
UnknownEntry	A query, update, or upsert operation failed because the state item does not exist.

Debug and Trace Messages

MSRS provides an interface for viewing error and informational messages returned by a service (refer to Figure 2-8). Filters allow you to restrict the types of messages displayed. For example, if you only select the Error check box, then only messages marked with a trace level of error will appear in the list. You can also restrict the output to display messages generated from a certain source or category.

To log the messages that appear in the console output service, you will need to call one of the following logging functions provided with the DSSP service base:

- **LogInfo** Logs message with a trace level of info. These messages will appear with two stars and a yellow background for the level column.
- **LogError** Logs messages with a trace level of error. These messages will appear with three stars and a pink background for the level column.
- **LogWarning** Logs messages with a trace level of warning. These messages will appear with no stars and no background color.
- **LogGroups** Logs categories and can be used when logging messages to assign them to certain categories. This allows you to restrict what messages are displayed in the Debug And Trace Messages window.
- **LogVerbose** Used to log messages generated from system services only.

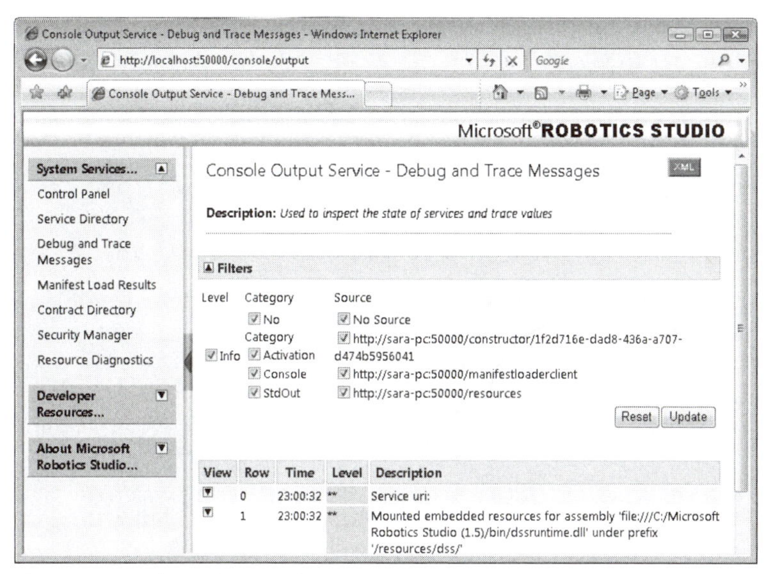

Figure 2-8 The Debug And Trace Messages window, which displays logged information from your service.

A trace level is used to indicate the type of message to log. DssInfo provides a way to log messages with a trace level of info. Messages logged as informational will appear with a yellow background in the level column (refer to Figure 2-8). You can log a message as informational by using code such as the following:

```
LogInfo("This is the state for my service " + _state.OutputMsg);
```

You can view details for a message by clicking the arrow that appears in the View column. This will display additional information such as the time the message was posted and from what service. For example, the detail for the informational message "This is information posted from my service Hello World" will appear as follows:

Message	This is information posted from my service Hello World
Category	StdOut
Level	Info
Time	2007-07-16T12:42:48.344-05:00
Subject	This is information posted from my service Hello World
Source	*http://sara-pc:50000/servicetutorial1*
CodeSite	Void Start()() at line:72, fileC:\Microsoft Robotics Studio (1.5)\samples\Service-Tutorials\Tutorial1\CSharp\ServiceTutorial1.cs

The other commonly logged message is an error. When performing functions that modify state, it is a good idea to include checks to ensure that the code updated successfully. If it did not, you can add code to post an error message along with the fault exception using the

LogError function. For example, the following code from Service Tutorial 3 will save the state only if it encounters no errors (otherwise, it will post the error message to the log):

```
yield return Arbiter.Choice(
   base.SaveState(_state),
   delegate(DefaultReplaceResponseType success) { },
   delegate(W3C.Soap.Fault fault)
   {
       LogError(null, "Unable to store state", fault);
   }
 );
```

The error message will appear in the Debug And Trace Messages window with a pink background and three stars as the level. The detail for the exception will include information associated with the exception and can be seen by clicking the arrow in the View column.

MSRS provides three trace levels: Informational, Warning, and Error. For service authors who wish to create their own trace levels, MSRS allows them to add new trace switches using the *System.Diagnostics.TraceSwitch* class. This class is part of the .NET Framework and is not specific to MSRS. You can get more information about creating new trace switches through the following URL: *http://msdn2.microsoft.com/en-us/library/system.diagnostics.traceswitch(vs.71).aspx*.

Summary

- Services are the basic building blocks for any application built with MSRS. Each service is used to accomplish a certain task, and several services can work together to accomplish a common goal. It is essential that these services can communicate. They do this by passing messages back and forth.
- Services are created by using a built-in Visual Studio template, which creates the basic class structure needed to support a service. Services are started using the command-line tool DssHost.exe.
- Services are like active documents, and the information they provide is known as the service's state. The value of the state variables changes, depending on when it is requested, and it typically contains information such as sensor readings.
- Service messages are sent and received through ports. A PortSet lists the operations that can be associated with a port and, thus, defines what types of messages the service can send and receive.
- Services communicate with each other through the use of partnerships. One service can subscribe to another and receive notifications whenever the state has changed. This can be useful when there is a need to monitor a robot's sensors or actuators.

Chapter 3

Visual Programming Language

Visual Programming Language (VPL) is a graphical application design tool that you can use to create and execute simple or complex robotics applications. The tool is intuitive and easy to use. You can drag and drop programming elements onto a design surface and then use connections to specify inputs and outputs (see Figure 3-1). The elements can be as simple as a data value or as complex as a service. This chapter will cover how to use the VPL design tool to write both a simple application and a custom activity. You will learn how to send notifications, use built-in services, and debug a VPL application. The chapter should be of interest to beginning and experienced developers.

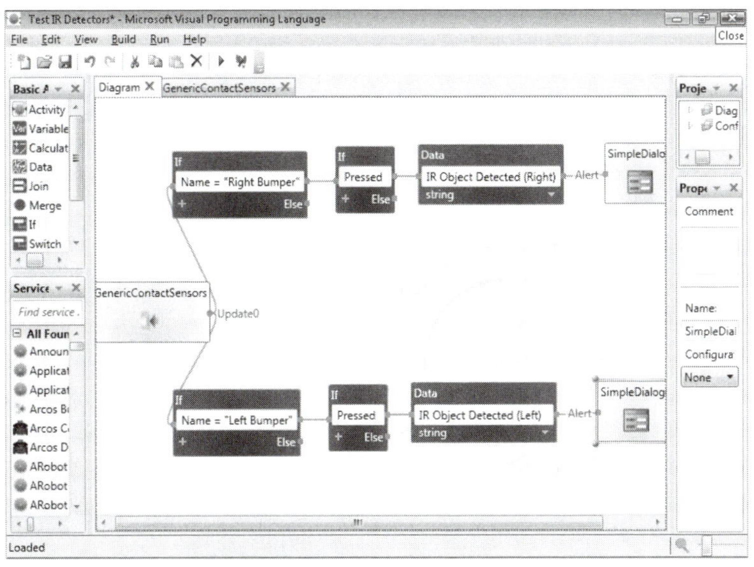

Figure 3-1 The VPL design tool features a drag-and-drop interface for creating robotics applications.

Overview of VPL

VPL differs from other application development tools in that it is data based and not control based, which affects how applications are executed. A typical Web application may involve users who initiate a sequence of commands based on some function they need to accomplish. For example, a Web application could be designed to place an order for a customer when the customer selects the items to buy and then initiates a checkout.

An application designed with VPL may involve several different processes operating one at a time or all at once. The applications are typically distributed, and operations may occur concurrently. For example, a robot may have several sensors that read information about the robot's environment. A bumper may indicate whether the robot has encountered another object. A motion sensor could indicate whether an object around the robot is moving, and a GPS device could give the robot's location. The application that monitors this robot must be able to receive data from all of these sensors concurrently. At the same time, it must then be able to issue commands, such as sending power to the motors that operate the robot's wheels. To accommodate this type of scenario, an application development tool must be able to reflect nonsequential operations.

Who Can Use VPL?

VPL can be used by both beginning and experienced programmers. So what constitutes a beginner? Even though VPL is easy to use, readers might feel quickly overwhelmed if they do not understand programming fundamentals such as variable assignments or conditional logic. The Microsoft Robotics Studio (MSRS) documentation claims that nonprogrammers can create robotics applications using VPL, but I am not so sure about this.

VPL is not an object-oriented language, and it was not designed to teach people how to program. It does not provide access to the .NET Framework, and it is not meant to be a general purpose language. However, if a reader has worked at least briefly with another high-level development language, VPL can offer a quick and easy way to bypass much of the plumbing involved with creating a robotics application. It can also offer a wonderful way to visualize the program as a functioning whole and not a series of class files.

VPL may be useful to experienced programmers who wish to build an application quickly using pre-existing services. A team of programmers may wish to build custom base services that can be used to operate one or more robots. These services, along with the generic services already provided with MSRS, could then be used by all of the team members to piece together an application using VPL. Services could be used by multiple robots, thereby making the use of some services a common function. The decision whether to use VPL or the Visual Studio template will depend mostly on programmer preference.

Tip Services are one of the basic building blocks used in applications built with VPL. If you are not familiar with services and have not read Chapter 2, "Understanding Services," you should do so before continuing with this chapter.

What Are Activities?

You build a VPL application by dragging and dropping blocks onto the design surface. Each block can represent a basic activity or a service (see Figure 3-2), such as the ones covered in

Chapter 2. Basic activities perform simple functions, such as storing the value of a variable or implementing an If...Else statement. You can refer to a list of basic activities in Table 3-1.

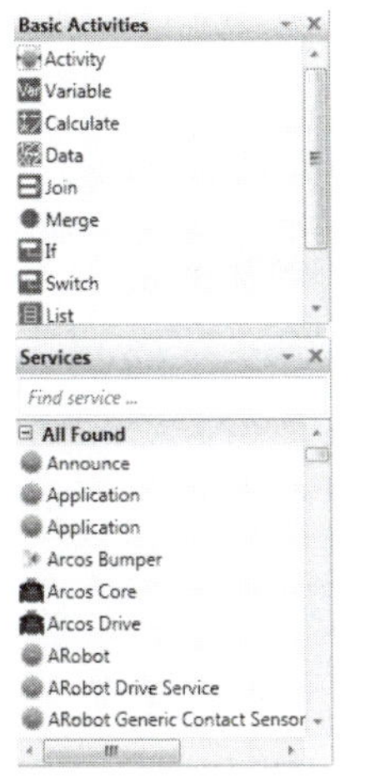

Figure 3-2 Basic activities and services are used to design a robotics application with VPL.

Table 3-1 List of Basic Activities Available in VPL

Name	Description
Activity	Used to create a new activity. The new activity is user defined and allows programmers to design their own custom activity blocks. Other basic activities and services can be used within the newly created activity. Each new activity can receive an input value and return an output value. The activity can also trigger a notification.
Variable	Used to store a value that can then be referenced in other activity blocks. You can use a variable to store the state from a service and then use that value across multiple activities or services. In this way, you use variables to share state between activities. The value of a variable can be any of the types supported by the .NET framework, such as *int*, *string*, or *Boolean*. It can also be a list containing one of the supported types. The variable names are case-sensitive, and the values can be set or retrieved.
Calculate	Used to perform simple arithmetic functions such as add, subtract, multiply, and divide. You can also use logical operators such as && for *and*, || for *or*, and ! for *not*. You can also use Calculate to access individual members of an input message. For example, you can use Calculate to access the different buttons on an Xbox 360 controller.
Data	Used to supply a data value for another activity box such as a Variable or Calculate. The data value can be one of the supported .NET data types.
Join	Used to combine the results from more than one activity block. By using this activity, you ensure that a message must be received from all the input connections before an output message is produced.
Merge	Used to merge the messages from two or more activity blocks. The resulting message is then sent on to the next activity block. This is different from a join, which will synchronize the messages.

Table 3-1 List of Basic Activities Available in VPL

Name	Description
If	Used to support a conditional expression such as an If..Then..Else statement. The expression can include operators such as = for *equals*, < for *less than*, and != for *not equals*.
Switch	Similar to a CASE statement, this activity block is used to match incoming values to a branch in the statement and route the message accordingly. For example, you could use a Switch block to determine which function was selected by a user.
List	Used to create a list of data items for a specific data type. The list may be used as input for another activity block.
List Functions	Used to perform a selected function such as append, sort, insert item, or concatenate on the items in an existing list and produce a new list as the result.
# Comment	Used to add descriptive text to an activity block.

The basic function of an activity is to accept an input message and produce an output message. What happens in between, or the action that is applied, depends on what type of activity it is. Activities are linked together through the messages passed between them. An output message from one activity becomes the input message for another activity.

The basic activity named Activity allows you to create a user-defined activity. This activity can contain other activity blocks and can even include a reference to itself. The activity can accept an incoming message and return an outgoing message. It can also support one or more notifications. The notifications allow an activity to generate multiple output messages, depending on the result of some user-defined action.

One of the really nice features about VPL is that it allows you to compile your application as a service. When this is done, all the activities within that application are included with the service. This allows you to reuse one activity within another VPL application–this time as a service. The activity you just compiled now appears in the list of services, and you can use it within another VPL application.

You can also use services as activity blocks in VPL applications. VPL allows you to use any Decentralized Software Services (DSS) service built with MSRS in your VPL application. This includes not only the built-in services provided with MSRS but any services you create yourself. Many of the services that come with MSRS are specific to a platform. For example, the following four services specifically support the Boe-Bot:

- **Boe-Bot BASIC Stamp 2** Provides control access to the BASIC Stamp 2 controller, which provides the brains for the Boe-Bot. This controller service is used to represent the Boe-Bot.
- **Boe-Bot Generic Contact Sensor** Provides access to the contact sensors supported by the Boe-Bot, which include whiskers and infrared detectors. These can be used as bumpers for the Boe-Bot and indicate when a collision is possible.

- **Boe-Bot Generic Drive** Uses the generic differential drive contract to implement drive functionality specific to the Boe-Bot. This allows you to drive the Boe-Bot by sending alternating power levels to the motors powering the left and right wheels.
- **Boe-Bot Generic Motor** Uses the generic motor contract to implement motor functionality specific to the Boe-Bot. This allows you to control whether the motors are enabled or disabled and allows you to set power levels for the motors.

To better understand all the different services available in MSRS, it is helpful to group them into categories. MSRS includes several generic services (see Table 3-2), which can be used to operate a variety of robotics platforms. You can incorporate these services into your VPL applications to perform tasks such as driving the robot and receiving feedback from the sensors. The services are located in the bottom left pane of VPL (see Figure 3-2, shown previously).

Some of the platform-specific services use these generic services as activities. For example, the service named Boe-Bot Generic Drive utilizes the Generic Differential Drive contract to partner with the Boe-Bot BASIC Stamp 2 service.

Table 3-2 Generic Services Included with MSRS and Available as Activities in VPL

Name	Description
Generic Analog Sensor	Provides access to an analog-type sensor, such as the light sensor included with the LEGO Mindstorms NXT.
Generic Analog Sensor Array	Provides access to an analog-type sensor array.
Generic Articulated Arm	Allows you to control an articulated, multi-jointed robotic arm, such as the KUKA LBR3.
Generic Battery	Used to provide access to a battery sensor, such as the one included with the LEGO Mindstorms NXT.
Generic Contact Sensors	Commonly used service that allows you to receive feedback from the robot's contact sensors. This can include sensors such as the whiskers included on the Boe-Bot or the bumpers included on the Create by iRobot.
Generic Differential Drive	Commonly used service that allows you to operate a robot with two wheels. The motors that drive each wheel must accept power values, and the difference between these values determines in which direction the robot moves. For example, if the two values are the same positive value, the robot moves forward in a straight line. If the power for the right motor is less than the power to the left, the robot will turn to the left. Negative values cause the motors to work in reverse.
Generic Encoder	Commonly used service that provides access to a DC motor with encoder feedback, such as the one used by the LEGO Mindstorms NXT.
Generic Motor	Commonly used service that is used to operate a servo motor. This allows you to enable and disable motors, as well as set power levels for all motors.
Generic Sonar	Used to operate a sonar sensor, such as the one used by the LEGO Mindstorms NXT.

Table 3-2 Generic Services Included with MSRS and Available as Activities in VPL

Name	Description
Generic SQL Store	Used to provide access to a Structured Query Language (SQL) database.
Generic Stream	Provides a bidirectional, packet-based stream access.
Generic WebCam	Used to provide access and capture images from a Web camera.

In addition to the generic services and platform-specific services, there are a handful of miscellaneous services (see Table 3-3) that demonstrate useful robotics behavior. For example, the Explorer service enables a robot with a differential drive system, bumpers, and laser range finder to explore its surroundings. You can access the code for these services in the MSRS installation directory within the \samples\misc subdirectory.

Note Keep in mind that any DSS service created with MSRS can be used in a VPL application. The services listed in Tables 3-2 and 3-3 are just some of the services you can use. Any service you create will also appear in the list of available services.

Table 3-3 Miscellaneous Services Included with MSRS and Accessible from VPL

Name	Description
Blob Tracker	Used to locate specific BLOBs or regions within an image for simple color tracking. A test service is also included.
SQL Client for ADO.NET	Provides database access to a SQL database using the ADO.NET data provider. This could be useful for permanent storage of state.
Direction Dialog	Implements a simple Windows-based dialog box that can be used to drive the robot forward, backward, left, and right.
Explorer	Used to demonstrate exploration behavior in a robot. When the robot encounters an object with one of its sensors, it will back up and turn the other way to avoid the object and explore the environment.
Follower	Used to demonstrate multiple behaviors such as text-to-speech, obstacle avoidance, and visual tracking.
Game Controller	Used to access a DirectInput game controller such as a joystick or gamepad.
Simple Dashboard	Used to access a Windows-based dialog box that allows you to perform various functions such as driving the robot continuously with a mouse.
Simple Vision	Provides access to a basic set of services that allow robots with a camera to detect color, faces, and hand gestures.
Syndication	Provides access to an ATOM/RSS syndication service that allows users to access and update blogs.
Windows Messages	Used to send and receive Microsoft Windows messages. Includes access to a test service.
XInput Controller	Provides access to a controller that can be used to control an Xbox 360 gamepad.

Using the VPL Design Tool

VPL is a graphical design tool in which you build applications by dragging and dropping activities from the left-hand panes onto the main design diagram. By default, VPL starts off by opening an empty diagram (see Figure 3-3). You create an application by dragging activities or services from the left-hand toolboxes onto the design surface.

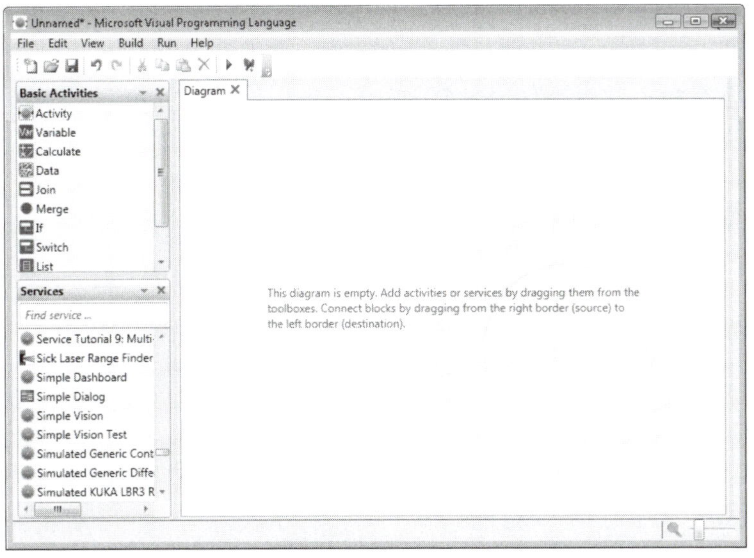

Figure 3-3 Empty VPL diagram.

Activities are linked together through the exchange of messages. Each activity applies an action to an input message and produces an output message. The output message from the source activity becomes the input message for the target activity. For example, assume you had a VPL diagram with two activities: a data activity and a variable activity. The data activity would provide the input for the variable activity. If the data activity was set with a value of false, the output message for the data activity would include the false value. The input message for the variable activity would also include the false value.

You use connections to facilitate the exchange of messages between activities. Connections are established between activities by clicking on the output of the source activity and dragging a line to the input of the target activity. An additional dialog box will then be displayed, which allows you to select an action from the target activity. When that action is applied to the input message, an output message is produced and sent to the target activity.

Establishing Connections

To better understand how connections work, let's consider the data and variable activities used earlier. It is good programming practice to initialize a variable with a data value. To do this in VPL, you would need to drag both a data and a variable activity onto the design surface.

To define a new variable, click the ellipses to open the Define Variables dialog box. Click Add and replace the field name with the variable name of your choice. You will also need to select a variable data type. In Figure 3-4, the variable is named DisplayMessage, and it will contain a data type of string.

Tip Keep in mind that VPL programming is not like conventional programming. To avoid unnecessary complexity, you should keep the creation of local variables to a minimum.

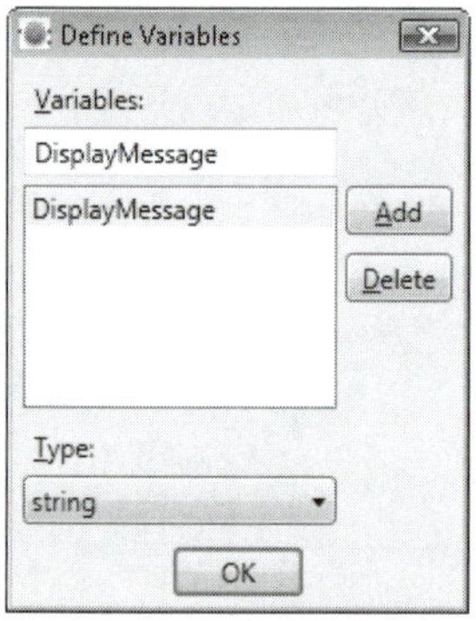

Figure 3-4 The Define Variables dialog box is used to define a new variable in a VPL application.

To set the value for a data activity, you can type the value directly into the data text box control. You will also need to select the appropriate data type from the drop-down list beneath the data text box control. The final step will be to establish a connection between the data and variable activity. To do this, drag a line between the right side of the data activity to the left side of the variable activity. The left side of a block represents the output message, and the right side is the input message. When you do this, a Connections dialog box (see Figure 3-5) will appear and prompt you to select the From and To values for this connection. In this case, there is only one From value available and, because you want to set the value of the variable, you should select SetValue as the To value. In this case, SetValue represents the action to be applied to the input message.

After the connection is established, the main diagram will display two blocks with a line drawn between them (see Figure 3-6). You could then build the remainder of the VPL program by dragging additional blocks onto the design surface.

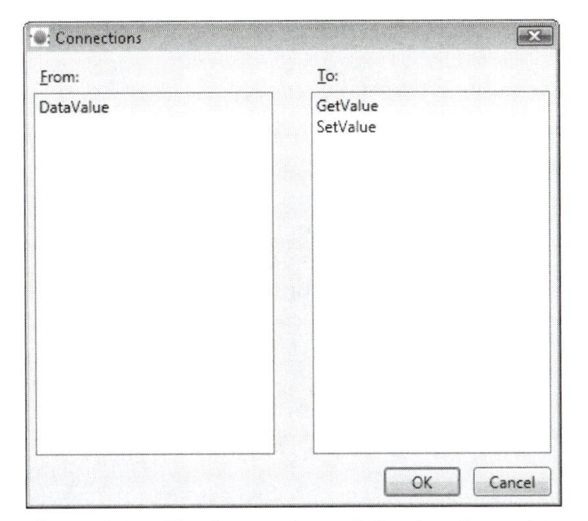

Figure 3-5 The Connections dialog box is used to select From and To values for a connection.

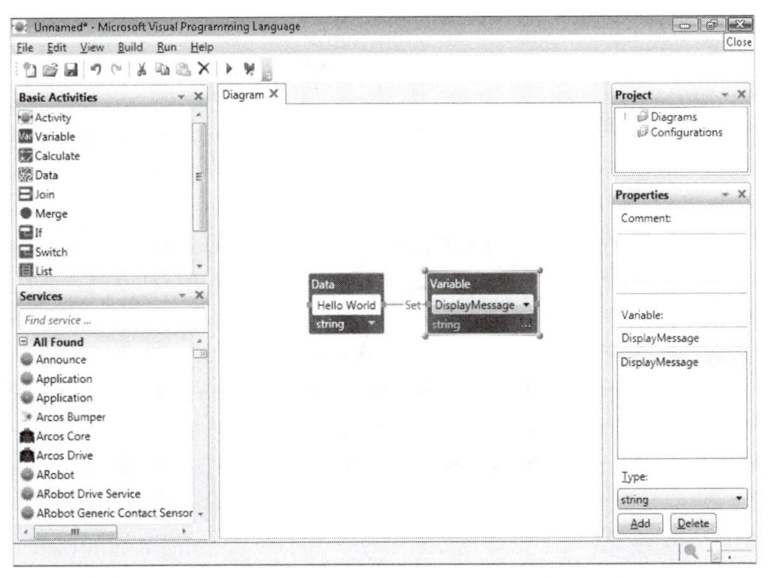

Figure 3-6 End result for a VPL diagram that is used to set the variable named DisplayMessage with the string value "Hello World."

Configuring a Service

When the activity is a service, you can double-click on the service to access the configuration. The Set Configuration page (see Figure 3-7) allows you to configure a service. The Set Configuration drop down box allows you to select one of the following four options:

- **Use a manifest** This allows you to specify an existing manifest file to use when loading the service. Once this option is selected, an additional button named Import Manifest will appear. You can click this button and select one of the applicable manifests from a list of available files.
- **Use another service** This allows you to specify the implementation for a specific service. This can be useful when designing VPL programs that use generic services. You can use the configuration option to specify a platform-specific service after the initial VPL program has been written.
- **Set initial configuration** This is used to specify initial state values for a service and can also be used to select and configure additional partner services. Partner services allow a service to take advantage of the functionality within another service. For more information about partner services, refer to the section titled "Communicating with Other Services" in Chapter 2, "Understanding Services."
- **None** In many cases, no special configuration is necessary, and this is a perfectly valid option. For example, this option works for the Simple Dialog and Text-To-Speech services.

Figure 3-7 shows the initial configuration options for the WebCam service. It is here that you would specify dynamic parameters such as the camera's device name and file name used to store the captured image. You can also specify the camera's image size, field of view angle, and compression quality level setting. If available, you can also set the initial configuration parameters. In this example for the WebCam service, the initial configuration parameters are used to specify the CameraDeviceName and CaptureFile.

The Set Configuration page includes two tabs: Partners and Initial State. You can associate one or more partners with a service, and each of these partners may be associated with a configuration file. Alternatively, a service may have no partners associated with it. When you specify the initial configuration options, VPL will include a configuration file with your VPL application.

Another way to configure services is by using a manifest file. The manifest is an XML-based file that contains a list of services that should be started together. This includes the configuration file, which will vary depending on the needs of the service. The configuration file is also an XML-based file that includes the dynamic configuration parameter values needed for the corresponding service.

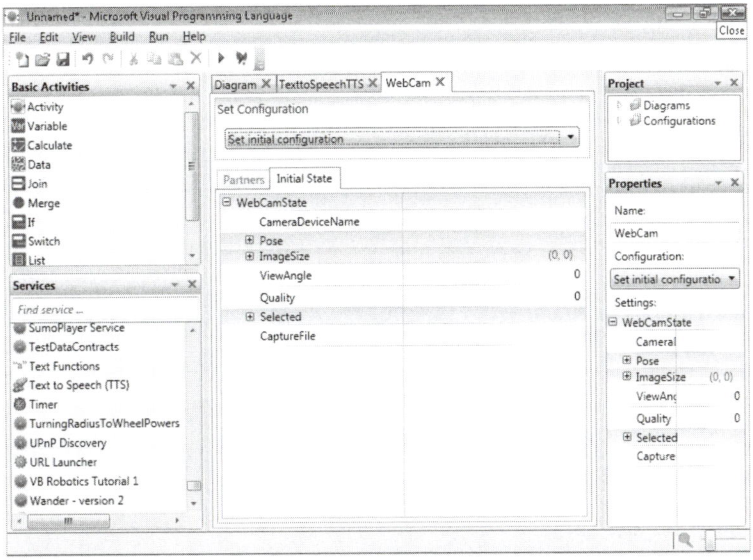

Figure 3-7 The Set Configuration page is used to configure a service using either a manifest or another service.

MSRS includes services and configuration files that can be used to control a handful of robots available for purchase. One such robot is the Boe-Bot by Parallax. When working with the Boe-Bot for an application designed with VPL, you have the option of using a manifest to specify configuration parameters for the Boe-Bot. You do this by selecting Use A Manifest from the drop-down list box and clicking Import Manifest. The Import Manifest dialog box (see Figure 3-8) will then list all manifest files that may be loaded for this service.

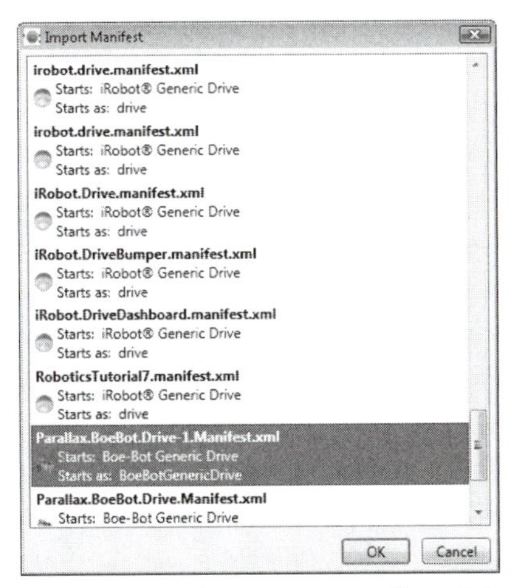

Figure 3-8 The Import Manifest dialog box is used to select a manifest that will be loaded with your service.

> **Tip** You may see more than one manifest file listed for a specific robot. If this is the case, you can use the mouse to hover over the entry. At this point, you should see a tooltip that shows you where the manifest file is located on your development machine. This should help you determine the correct manifest file to use.

If you choose to import the manifest, you will need to first edit the configuration file and change the value for the serial port element. The configuration file may include elements that are used to store state values. For example, the configuration file used to start the Boe-Bot looks like the following:

```
<Configuration>
    <Delay>0</Delay>
    <SerialPort>5</SerialPort>
  </Configuration>
  <AutonomousMode>false</AutonomousMode>
  <Connected>false</Connected>
  <FrameCounter>0</FrameCounter>
  <ConnectAttempts>0</ConnectAttempts>
  <MotorSpeed>
    <LeftSpeed>0</LeftSpeed>
    <RightSpeed>0</RightSpeed>
  </MotorSpeed>
  <Sensors>
    <IRLeft>false</IRLeft>
    <IRRight>false</IRRight>
    <WhiskerLeft>false</WhiskerLeft>
    <WhiskerRight>false</WhiskerRight>
  </Sensors>
```

Elements such as *Connected*, which specifies if the BASIC stamp is currently connected, represent items saved to state by the BasicStamp2 service. The BasicStamp2 service is provided with MSRS and is one of the services that allow you to control the Boe-Bot. In the next section, you will learn what specifically needs to be done to configure and work with the Boe-Bot robot.

A better way to configure your VPL services is to use VPL directly. For those services that expose their state, you could select Set Initial Configuration from the drop-down box. This will display the Initial State for that particular service. This includes all the state variables, and, from here, you can enter a value for variables such as the COM port.

Working with the Boe-Bot

At this point, I am sure you are anxious to get started working with an actual robot. After all, it is not enough for us programmers to write code; we want to see the results of our efforts moving through space. The sample code in this chapter will work with Parallax's Boe-Bot. (You can get more information about this robot through the following URL: *http://www.parallax.com/detail.asp?product_id=21832*.) The Parallax Web site offers a Boe-Bot kit that works with MSRS. This kit includes a Bluetooth wireless module and allows you to assemble your own Boe-Bot in a few hours using only basic hand tools.

The Boe-Bot is a small but sturdy robot (see Figure 3-9) that is expandable and easy to put together. It is very popular with students and hobbyists, and there are more than 90,000 Boe-Bots being used throughout the world. It is one of the robots supported by MSRS and provides an excellent opportunity for readers to get acquainted with how robots work.

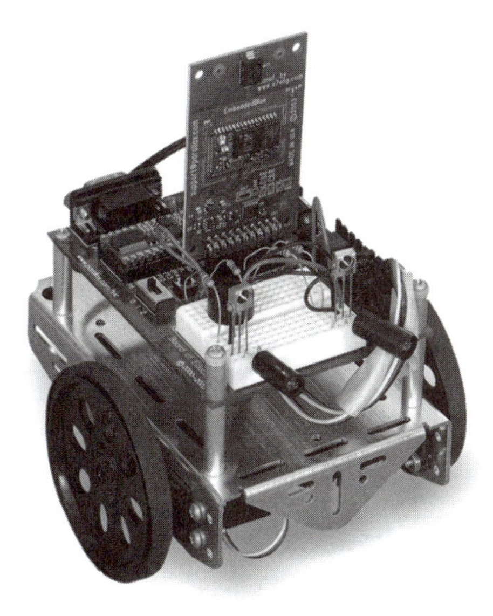

Figure 3-9 Parallax's Boe-Bot is one of the robots supported by MSRS and featured in this book.

The Boe-Bot kit comes with a manual that states it will take you 40 hours to complete the assembly. However, it should only take you two to three hours, and you do not need to read the entire manual or install all the sensors before returning to this chapter. At a bare minimum, you will need to read and complete the following sections (from the Boe-Bot manual) before attempting to work with the Boe-Bot:

- Chapter 1 – Activities 1, 2, 3, and 4
- Chapter 2 – Activities 3, 4, and 6
- Chapter 3 – Activities 1, 2, and 3
- Appendix D

Optionally, you can install the contact sensors that come with the Boe-Bot. This includes a set of whiskers and infrared (IR) detectors. To install these, you will need to review Chapters 5, 7, and 8 from the Boe-Bot manual. These sensors, which are used for object detection and distance measurement, will be referenced in the remaining sections of this chapter. If you prefer, you can install only one of the contact sensors, such as the whiskers and not the IR detectors.

Alternatively, you can install both sets of contact sensors, which enables your Boe-Bot to operate more effectively when navigating its environment.

After you have assembled the Boe-Bot, you will need to configure it to work with MSRS. This will involve downloading a control program to the Boe-Bot, which acts as a middle layer between the Boe-Bot and MSRS. It will also involve configuring the Boe-Bot for Bluetooth capabilities. For more information about the steps needed to configure your Boe-Bot, refer to Appendix B, "Configuring Hardware," the section titled "Configuring the Boe-Bot."

Writing a Simple VPL Application

After you have built and configured your Boe-Bot , you can start writing VPL applications to operate it. In this section, we will examine the steps necessary to build and execute a simple VPL application named BoeBotBumperTest. This application will be responsible for instructing your development machine to say "An object was detected" each time one of the IR detectors or whiskers is triggered. The IR detectors are triggered when an object passes within a few inches of the left or right IR detectors. The whiskers are triggered when one of them connects with the pins located on the Boe-Bot's breadboard. If you have not already installed one of the contact sensor sets, you may wish to do so at this time.

> **Tip** While finalizing the assembly of your Boe-Bot, you may go back and forth between using the serial connection to execute test programs directly on the robot and using the wireless module to execute programs remotely with MSRS. If you do this, do not forget to reload the BoeBotControlForMsrsCtp2.bs2 program before you return to using MSRS. This driver program is what enables MSRS to communicate directly with the robot and must be the last program that is loaded onto the Boe-Bot.

To begin, you need to open the VPL tool by clicking Start, All Programs, Microsoft Robotics Studio (1.5), and then Visual Programming Language, which initiates a new VPL diagram. You can use the Find Service text box at the top of the Services toolbox, or you can scroll through the list until you locate the Generic Contact Sensors service. Drag this service onto the diagram surface. The Generic Contact Sensors service provides access to base functions that can be used by multiple robotics platforms.

To establish a link between the Generic Contact Sensors service and the Boe-Bot, you will need to configure the sensors service. You do this by double-clicking the service block in the main diagram. This opens the Set Configuration page where you can select Use A Manifest from the drop-down list box, and then click Import Manifest. The Import Manifest dialog box displays a list of manifest files supported by this service. You need to select the Parallax.MotorIrBumper.manifest.xml manifest and then click OK to tie this service to the Boe-Bot. It is this manifest file that informs the sensors service about which sensor should be monitored. By default, this manifest file is located in the \samples\config folder for your local MSRS installation.

Remember, the Boe-Bot is not the only robot that you can use with the Generic Contact Sensors service. The Import Manifest dialog box contains a list of manifest files provided by other robotics hardware manufacturers. This list includes the following:

- iRobot's Create
- LEGO's NXT and RCX
- MobileRobots's Pioneer 3DX
- fischertechnik's Bionic Walker

> **Note** It is possible for the generic services to work with more robots than the ones listed here. For example, you may download services for another robot from the robot manufacturer's Web site. Alternatively, you could create your own set of services for a custom-made robot. These custom-made robots can also potentially work with the generic services. If manifest files exist for these robots and the robots implement the generic services, they will also be listed in the Import Manifest list.

After the Generic Contact Sensors service is configured, you can return to the diagram tab by clicking on the tab at the top of the design surface. You will then need to drag and drop the data activity from the Basic Activities toolbox. Place the data activity to the right of the Generic Contact Sensors service. Click the text box inside the data activity block, select the 0 that is displayed by default, and type **An object was detected**. Next, select string as the data type.

To connect the two blocks, use your mouse to drag a line from the round Notification connector on the right side of the GenericContactSensors block to the input block on the left side of the Data block (see Figure 3-10). This action initiates the Connections dialog box and prompts you to select a From and To event. By default, the To event is Create. Click OK to close the Connections dialog box. Because you need to know when the IR sensor has detected an object, select ContactSensorUpdated as the From event.

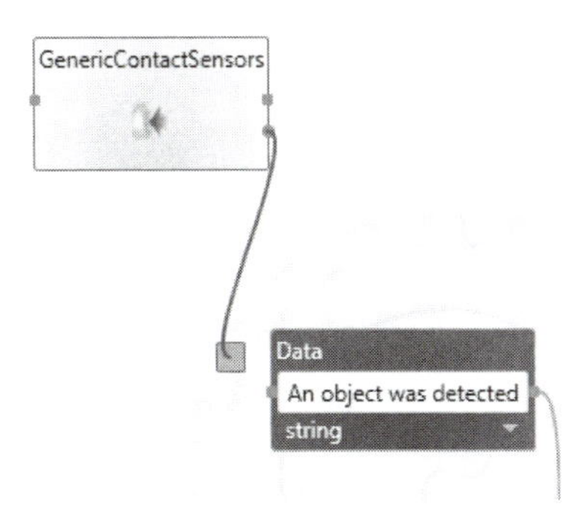

Figure 3-10 To connect the GenericContactSensor and Data blocks, drag a connection between the round Notification connector and the left-hand side of the Data block.

Your next step is to drag the Text-to-Speech service onto the diagram and to the right of the Data block. This service supports several functions related to rendering text as speech through a Speech Application Programming Interface (SAPI)–compatible engine. A SAPI-compatible engine should already be located on your Windows development machine because the Microsoft Speech API version 5 has been available since the year 2000. If this is not the case, and you are encountering errors when trying to run programs using the text-to-speech service, you can download it free of charge from the Microsoft Web site: *http://www.microsoft.com/downloads*.

Note MSRS includes services that provide speech recognition and speech synthesis capabilities for both a SAPI-compatible engine and the .NET 3.0 System.Speech library. The service we use in this chapter uses a SAPI-compatible engine and not the .NET 3.0 library.

To connect the Data block to the Text-to-Speech service, drag a connector from the right side of the Data block to the left side of the Text-to-Speech block. This opens the Connections dialog box and prompts you to select a From and To value. Because you want to hear the text spoken, select SayText as the To value, and click OK to initiate the Data Connections dialog box (see Figure 3-11). This prompts you to select a value that will be set for the SpeechText variable. By default, the value selected is null. You must select Value from the drop-down list box; otherwise, no text will be spoken when you run the program. Click OK to close the Data Connections dialog box.

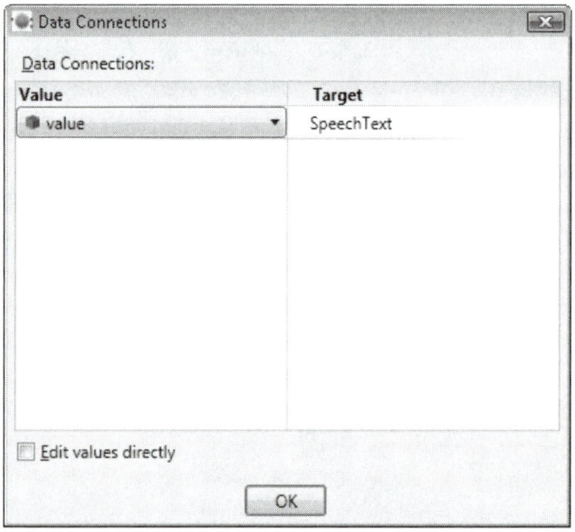

Figure 3-11 Use the Data Connections dialog box to select the value that will be set for the SpeechText variable.

> **Tip** If you forgot to select Value in the Data Connections dialog box, you can return to this dialog box by right-clicking the connector in the main diagram and selecting Data Connections. Alternatively, you can select the connection and use the Properties window on the right side of the screen to change the property assignment.

The *SayText* function will result in the text being spoken asynchronously or in parallel with all other processes called by the MSRS runtime. Alternatively, you could select SayTextSynchronously if you wanted to ensure that no other processes could execute while the text is spoken. Assuming the sound is enabled and you have desktop speakers installed on your development machine, you should hear the phrase "An object was detected" each time you pass your hand in front of the IR detectors or push on one of the whiskers.

> **Tip** You can find additional VPL examples in the Introductory Courseware available as a free download from the MSRS Web site. These helpful tutorials demonstrate how to perform complex tasks using VPL and a Create robot. Readers interested in these tutorials can download them from the following URL: *http://www.microsoft.com/downloads/details.aspx?familyid=f294c8e7-6617-4dd8-8354-7e97f3167e1a*.

Executing and Debugging a VPL Application

When you have completed the previous tasks, your VPL application should appear similar to the diagram in Figure 3-12. You can save the application as a file with an .mvpl file extension. It is best for you to save the application in a new subfolder within the MSRS application directory. You should name the application BoeBotBumperTest.mvpl. You can also run the application in a way that is similar to running a Visual Studio application. This includes the ability to step through your program using debug mode.

To execute the application without going into debug mode, click Start from the Run menu, or press F5. Alternatively, you could click the green arrow icon from the top left menu bar. This initiates the Run dialog box (see Figure 3-13). Because the DSS runtime is required even when the application is executing within VPL, a connection to the DSS runtime will be initiated.

> **Tip** By default, VPL will start the DSS runtime using 50000 as the HTTP port and 50001 as the TCP port. If you have chosen to work with different port numbers, you can change the values that VPL uses by clicking Port Settings from the Run menu.

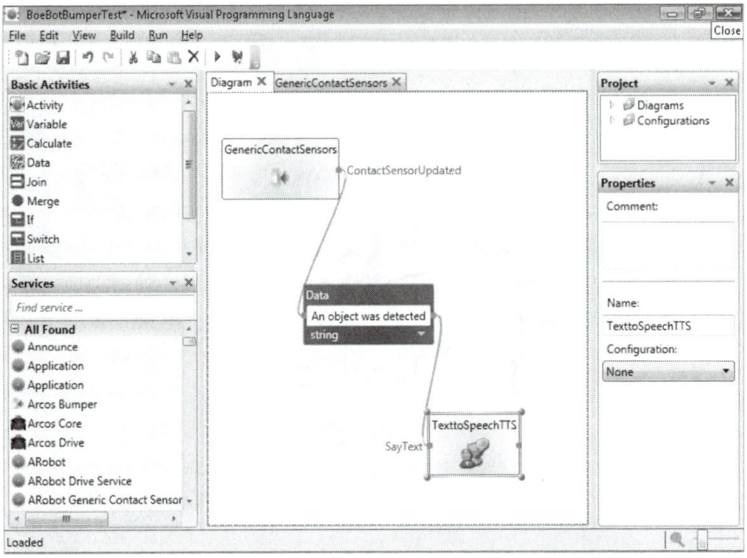

Figure 3-12 Complete diagram for the simple application design with VPL.

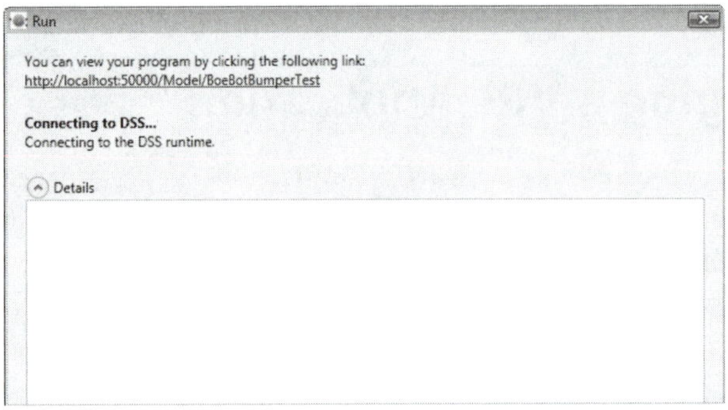

Figure 3-13 The Run dialog box automatically appears when you start an application in VPL.

After the connection is established, the application starts. This process could take several seconds, and you will know the Boe-Bot is ready when the robot's speaker beeps twice and the green light on the Bluetooth module turns on. At this point, you can trigger the contact sensors on the Boe-Bot to test the application.

Even if you chose not to run the application in debug mode, you can view the program by clicking the application link within the Run dialog box. For the BoeBotBumperTest application, the link should be *http://localhost:50000/Model/BoeBotBumperTest*. Clicking this link brings you to the debug view, but it will not allow you to step through the diagram like you can when you are in debug mode.

To run the application in debug mode, you need to select Debug Start from the Run menu, or press F10. This also initiates the Run dialog box, but when you click the application link, you can step through the VPL application using the Microsoft VPL Debug View (see Figure 3-14).

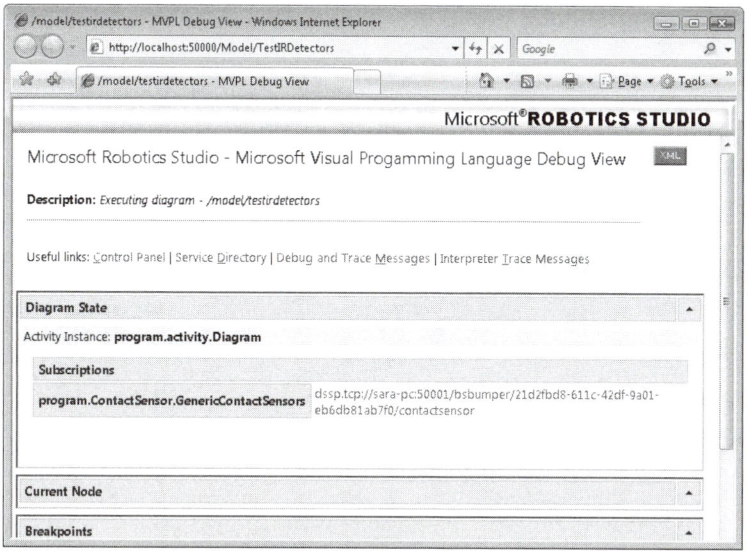

Figure 3-14 The Microsoft VPL Debug View allows you to step through a VPL program, set breakpoints, and view the values of state variables.

The MVPL Debug View, which is hosted inside of your Internet browser, includes the following four sections:

- **Diagram State** Displays the state of the current program and lists any partner subscriptions that have been established. For the BoeBotBumperTest program, a subscription has been created for the GenericContactSensors service.
- **Current Node** Displays the VPL diagram and indicates the status of the program. Arrow buttons allow you to step through the blocks and observe the values of state variables as the program is executing. For example, Figure 3-15 displays the value of state variables while executing the BoeBotBumperTest program.

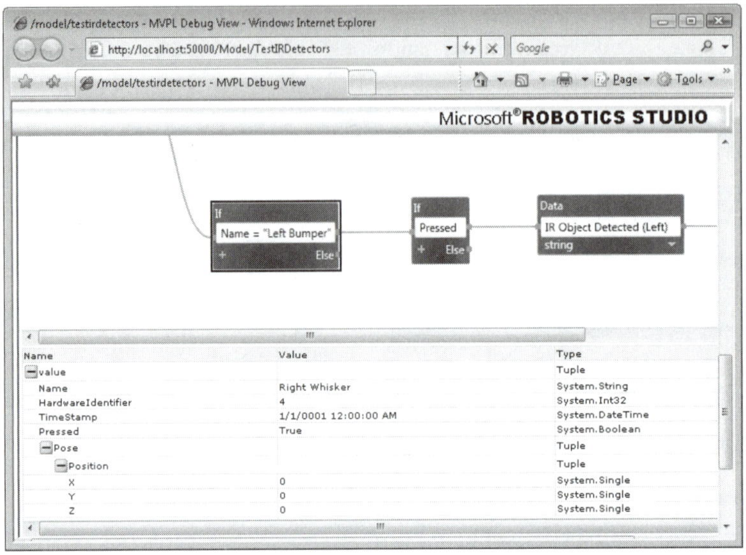

Figure 3-15 While in debug mode, you can step through the program and observe the values of state variables as the program is executing.

- **Breakpoints** You can set breakpoints through the Pending Nodes section that appears in this section. This is particularly useful when you are debugging a VPL program that contains several blocks. You can clear, enable, or disable breakpoints from the Breakpoints section.
- **Pending Node** Lists all pending nodes for the current instance. From here you can set breakpoints by clicking the SetBP button located next to the desired node.

Hardware identifier values are used to distinguish each contact sensor. The services that are used to operate the Boe-Bot automatically assign hardware identifier values 1 through 4 (see Table 3-4).

Table 3-4 **Hardware Identifier Values Assigned to the Contact Sensors for the Boe-Bot**

Sensor Name	Hardware Identifier
Left IR Bumper	1
Right IR Bumper	2
Left Whisker	3
Right Whisker	4

Compiling as a Service

One of the nicer features of VPL is its ability to compile your application as a service. Unfortunately, there is no way to go the other way and turn a service into a VPL application. After you compile a VPL application as a service, you can select it from the list of services in the left-hand menu. This can be useful when you need to group together commonly used blocks into a function that you can reuse in other applications.

To compile the application as a service, click Compile As A Service from the Build menu. This creates a Visual C# project, which you can open in Visual Studio to view the code associated with your VPL application. This can be very useful for programmers who are new to MSRS or programming in general. It provides an opportunity to understand how blocks in VPL relate to the code required to build a service.

> **Note** When you compile the application as a service, by default, it places the code files for your service on your desktop.

For example, if we compile the BoeBotBumperTest as a service, it creates a Visual Studio project that contains the following two class files:

- **DiagramService.cs** This is the implementation class, and it is where the code that runs your application is located. It is where the message handlers responsible for generating notifications to subscribers are located.
- **DiagramTypes.cs** This is where the type definitions, such as the ones that handle state, are located. It is also where the PortSet is located. The PortSet is used to define what types of Decentralized Software Services Protocol (DSSP) operations are allowed.

If you open the project in Visual Studio and then expand the References folder in Solution Explorer (see Figure 3-16), you see references to the Robotics.Common.Proxy and TextToSpeech.Y2006.M05.Proxy assemblies. These are the proxy files that handle calls to the two services included in the BoeBotBumperTest application. The Robotics.Common.Proxy assembly contains classes for all the generic services (refer to Table 3-2) available with MSRS. As you may recall, the BoeBotBumperTest application included a block for the GenericContactSensors service.

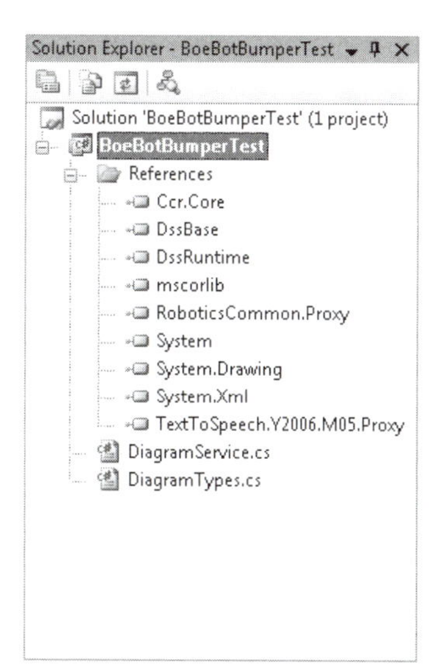

Figure 3-16 The References for the BoeBotBumperTest service include the *Robotics.Common.Proxy* and *TextToSpeech.Y2006.M05.Proxy* assemblies.

Examining the Generated Code

Even if you are not interested in using VPL as a design tool, you may want to consider using it as a learning tool. It can save you a lot of time by generating the code to interface with built-in services. For example, the BoeBotBumperTest service includes several class files related to handling what happens when one of the contact sensors is triggered.

Examining the code generated for the BoeBotBumperTest service can be useful for readers who want to learn more about the fundamentals of Concurrency and Coordination Runtime (CCR) and how to work with the GenericContactSensors service. For example, the message handler for the GenericContactSensors subscription uses the Task interface (ccr.ITask). This allows the code within the handler to execute in the context of the CCR dispatcher. The dispatcher manages threads for the operating system and ensures that tasks are load balanced properly.

```
IEnumerator<ccr.ITask>
   GenericContactSensorsUpdateHandler(contactsensor.Update message)
{
     OnGenericContactSensorsUpdateHandler handler = new
OnGenericContactSensorsUpdateHandler(this, Environment.TaskQueue);
     return handler.RunHandler(message);
}
```

You can then look at the code within the constructor for the *OnGenericContactSensorsUpdateHandler* class. The following code is what this constructor looks like:

```
     public OnGenericContactSensorsUpdateHandler(DiagramService service,
         ccr.DispatcherQueue queue) : base(service, queue)
     {

        // Register joins with base class to manage active count for
        // incomplete joins.
        base.RegisterJoin(_joinAlphaPorts);
        // Activate Join handlers
        Activate(ccr.Arbiter.MultiplePortReceive(true, _joinAlphaPorts,
            _joinAlphaHandler));
      }
```

Notice that the update handler is assigned to the DispatcherQueue. This means that the task executes within a First In, First Out (FIFO) queue of tasks. In this case, using CCR task scheduling to handle incoming messages is advisable because multiple sensors (whiskers and IR detectors) could be returning results at the same time that commands need to be sent to the Boe-Bot. All of this is happening very quickly, and it is the responsibility of the CCR dispatcher to properly manage all of these tasks.

Create a Custom Activity

VPL allows users to create custom activities, which you build by dragging blocks onto the activity design surface. To create a new activity, move an instance of the basic activity named Activity onto the main diagram. You can then double-click this activity to get to the activity design surface.

The activity design surface looks slightly different from the main diagram (see Figure 3-17). The square in the left border represents the input, and the square in the right border is the output. The circle in the right border relates to notifications, and an activity can have none, one, or several notifications associated with it.

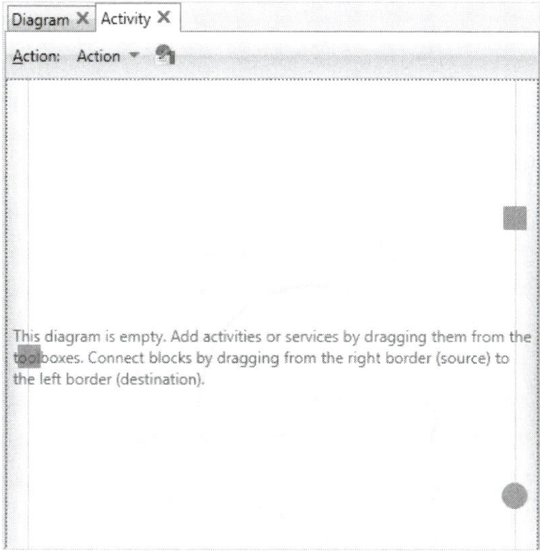

Figure 3-17 The activity design surface includes orange squares and a circle to represent the input and output data values as well as notifications.

You define one or more actions for each activity. For example, you could have an Add action and an Update action for the same activity. You access the Actions And Notifications dialog box (see Figure 3-18) by clicking the icon next to the Action drop-down list box. Actions are user-defined, and you can name them whatever you prefer. Each action can have one or more input and output values associated with it. The action is selected when you connect the activity to another block in the main diagram.

Actions and Notifications

Actions | Notifications

Actions: Add

Description: Used to add one number to the InVal and return as a string

Input values: InVal

Output values: OutVal

Type:

Type:

Add | Delete | Add | Delete | Add | Delete

OK | Cancel

Figure 3-18 Use the Actions And Notifications dialog box to define input and output values associated with an activity's user-defined action.

To demonstrate how this might work, consider a VPL program that simply counts to 10,000 and then logs that number as a trace message, as shown in Figure 3-19. This program is an example of a program that performs a loop.

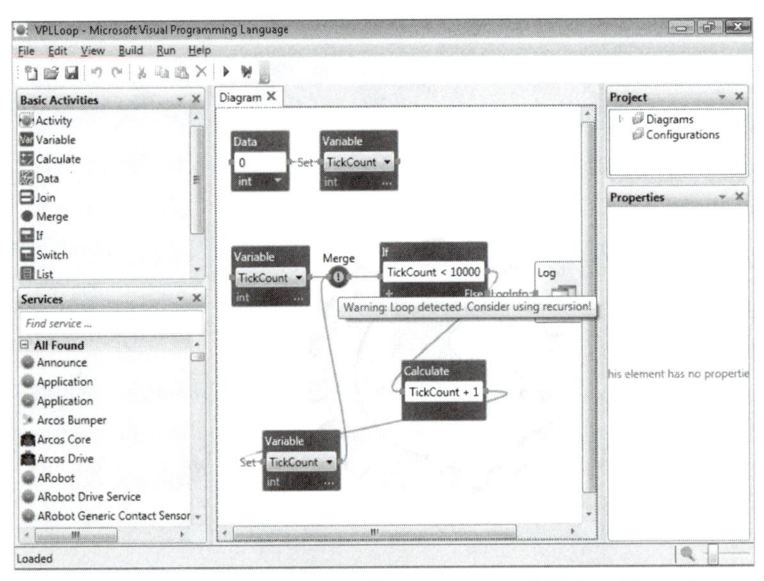

Figure 3-19 Simple VPL program that counts to 10,000 and logs the result as a trace message.

A better way of performing this same task would be to use a concept known as *recursion*. Instead of using a loop, in which a block loops back and references itself, you could use a user-defined activity. Recursion is an alternative to looping, and it generally results in more readable code. For example, to replace the VPL program shown in Figure 3-19, you could create a user-defined activity named Increment. The block for the new activity would exist not only in the main program, but there would be a reference to it within the user-defined activity as well (see Figure 3-20).

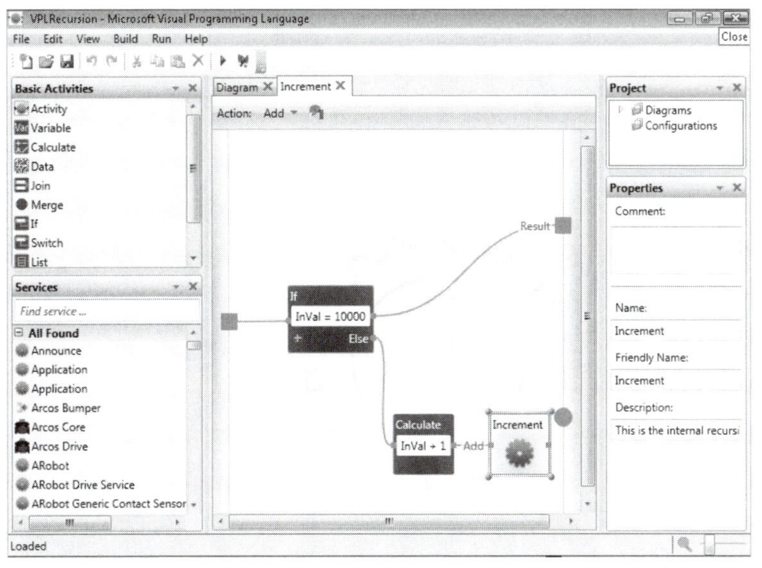

Figure 3-20 User-defined activity named Increment. This activity was created to implement a process known as recursion.

> **Tip** When implementing loops or recursive activities, it is a good idea to run the program in debug mode. This ensures that you are able to step through the program and to ensure that no infinite loops exist. An infinite loop could potentially cause problems for your development machine.

Summary

- Even though VPL is easy to use, it is still a powerful development tool in the MSRS arsenal. Users can drag blocks onto the diagram and link them together with connectors. Blocks can represent basic activities or services.
- Every activity is used to process messages. An outgoing message for one activity becomes the input message for another activity. Basic activities include simple constructs such as storage of a data value, a conditional statement, or the calculation of a value.
- Services can represent either the built-in services available with MSRS or services the users create themselves.Any DSS service created with MSRS can be used in a VPL program.

- This chapter featured the assembly and creation of an MSRS-supported robot named the Boe-Bot. This three-wheeled robot supports a Bluetooth module, which allows users to operate the Boe-Bot wirelessly.
- Using generic services such as the Generic Contact Sensors service, you can build a simple VPL program that can handle notifications whenever one of the contact sensors on the Boe-Bot is pressed.
- You can use the code generation tool within VPL to compile your VPL application as a service. This offers a valuable look at how you can use CCR to manage incoming tasks.

Chapter 4
Simulation

First off, let's briefly define what a simulation is. In regard to Microsoft Robotics Studio (MSRS), a simulation is a mathematical computer model that is used to represent one or more physical objects. Data is fed into the model, and the result may or may not be rendered to create a visual three-dimensional (3D) image. Sounds complicated? Well, in some ways, it is. But if you have played any recent software games, then you already have experience using a simulation. Today's high-powered video games are nothing more than elaborate simulations involving characters that the player is able to control. This chapter will assume that, although you may have played a game or ran a simulation, you have not actually created one yourself.

If you skimmed through Chapter 2 and thought none of it was very important, think again. In fact, you might want to go back and review it so you will not get lost in this chapter and the chapters that follow. Even though this chapter deals with a visual tool, it is not similar to the chapter on Visual Programming Language (VPL). VPL primarily involves dragging and dropping blocks onto a design surface. This greatly simplifies the process of creating a robotics application because the majority of code is generated automatically through the tool.

The main component of simulation is not a tool but rather an engine; an even more accurate description is that the simulation engine is a service. To create a simulation, you first create a Decentralized Software Services (DSS) service (like the ones created in Chapter 2), and this service will use the simulation engine as a partner service. This chapter will cover how to create a simulation and work with entities. This will include working with the graphical editor provided with MSRS and also creating a simulation using a DSS service.

Why Create Simulations?

The biggest benefit to using a simulation tool for robotics is that everything is virtual, and you do not need to have an actual robot. For programmers working with expensive robots, this can be a huge advantage. This is especially true when work is done by a group of programmers who each need time with the robot. With the simulation engine, each programmer can create his or her own simulation service and work with the robot independently. The programmer can carry out experiments and see how the robot reacts to different scenarios.

This can also be an advantage for students or hobbyists first learning about robotics. In the past few years, there has been a shift toward including robotics classes, not only in colleges, but at the high school level. Not all of these educational institutions can afford a robot for every student. The institution might be forced to work with a simple robot just because it is more affordable, but it may not be what the instructor would prefer. By using the simulation engine, students can work with a wider variety of robots.

One last benefit to consider is the opportunity to prototype a new robot design or experiment with a new scenario. Robots can be very expensive, and, rather than risk damaging the robot, a programmer can first experiment with a simulation rather than working with the actual robot. Some of today's robot competitions require robots to navigate through a complex obstacle course. These obstacle courses can involve multiple levels, and robots can fall off platforms or run into objects or other robots. The simulation engine can be used to design a navigation path for the robot before it attempts to navigate the actual course. After the code has been stabilized, a simple manifest change will enable a programmer to use the same code with an actual robot.

Now, I would be remiss if I did not mention the drawback to working with a simulation. Oddly enough, it is the same reason that it is a benefit. Because the simulation is virtual and does not involve a robot moving about in the real world, it is not always representative of an actual environment. The real world is messy and full of unknown variables. A simulation involves predictability with known constraints, and this is not always the case when working with real robots. Consider, for example, an infrared detector, which robots commonly use to measure distance. This simple sensor can easily be triggered by fluorescent lighting in a room. Therefore, when bringing a robot to a new environment, it is advisable to first test the sensor within the room to ensure that it is not triggered by an overhead light. This is the type of noisy data that simulations do not take into account.

Using the Visual Simulation Environment

Even though the simulation engine is just a service, MSRS provides an editor that you can use to create new entities and arrange them within a simulation scene. The graphical editor—named the Visual Simulation Environment (VSE)—operates in two basic modes: run and edit. Run mode will run the simulation but does not allow you to change any of the properties affecting the simulation scene. Edit mode (see Figure 4-1) allows you to alter the current scene by changing the properties of existing entities or adding new entities.

This section will include information you need to work with VSE. Because simulation is a heavily graphics-intensive process, you will need to make certain hardware considerations before starting out with VSE. You will also need to understand entities and know how to create new entities.

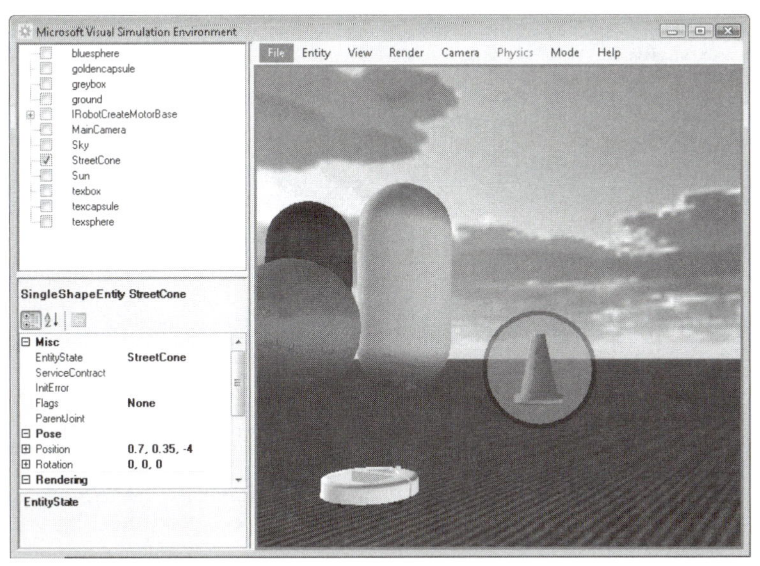

Figure 4-1 Simulation scenes can be modified and new entities added while in edit mode.

Hardware Considerations

To render simulations and use VSE, your development machine must include a compatible graphics card. The graphics card must support pixel and vertex shaders, but most laptops or desktop machines that are capable of running current graphics-based games will do this. Specifically, your graphics card must support the vertex shader VS_2_0 or higher and pixel shader PS_2_0 or higher. This means that the card must be compatible for DirectX version 9 or higher.

If you are not sure which of these shaders your graphics card uses and you have already installed MSRS, you can open one of the VSEs, such as the Basic Simulation Environment. Next, click on the Help menu, and then select About Microsoft Visual Simulation Environment (see Figure 4-2). If you scroll to the bottom of this dialog box, you will see what pixel and vertex shader versions your graphics card supports.

Tip The following URL lists graphics cards that have been tested and approved by the MSRS team: *http://channel9.msdn.com/wiki/default.aspx/Channel9.SimulationFAQ*. Although this list does not include all graphics cards that may support VSE, it does cover some of the more commonly used ones. You can also check the following Web site to determine what pixel and vertex shaders your graphics card supports: *http://www.techpowerup.com/gpudb/*.

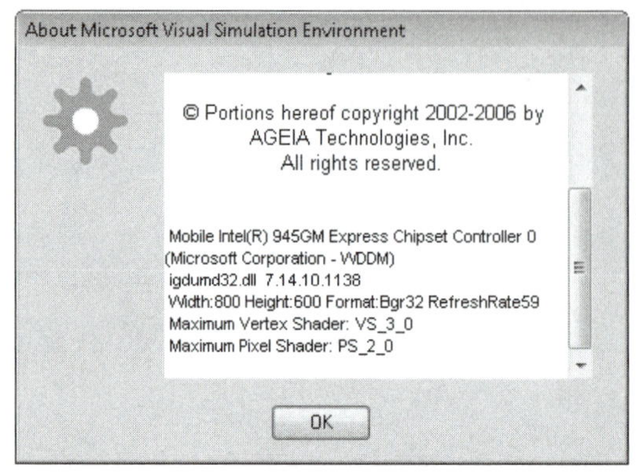

Figure 4-2 The About Microsoft Visual Simulation Environment dialog box displays information about which vertex and pixel shader versions the development machine is using.

It is possible to install MSRS on a development machine, work with services and the robotics tutorials, and still not be able to access VSE. One way to tell if your graphics card supports simulation is to install MSRS and try to run one of the simulations provided. If only the frame and menu bar of the VSE window appear, your card does not support the DirectX 9 shaders. You will either have to upgrade your graphics card or use a different development machine that does have a compatible graphics card.

It is also possible that your graphics card will not support the required shaders but will still try to render the scene. This is because there are two ways to perform graphics functions: by using software or by using hardware. If the functions are performed by the hardware, then the graphic will render very quickly. If the hardware does not support the graphics function, then the simulation will run in software mode. In general, software performs much more slowly than the hardware, and thus you may see significant performance degradation. To find out which shader versions your graphics card supports, refer to About Microsoft Visual Simulation Environment from the Help menu.

Using the AGEIA PhysX Card

MSRS uses the AGEIA PhysX engine to provide physics for the simulation environment. The MSRS installation contains the installation for the AGEIA PhysX engine, but you can optionally purchase and install the add-in hardware physics processor made by AGEIA. Although the AGEIA Web site (*http://www.ageia.com*) lists the price of the PhysX processor as $299 U.S. dollars, you can find it for much cheaper at local retailers such as Radio Shack or Fry's Electronics. You can also buy it online through several retailers. I was able to buy a card for only $145 U.S. dollars through TigerDirect.com.

> **Tip** Depending on the complexity of the simulation you are running, you may not notice any differences in the simulation itself. Before running out and buying a PhysX card, I would suggest you work with the simulation tool and see if it meets your needs without the additional hardware purchase.

If you decide to purchase and install the card, keep the following in mind. Prior to installing the PhysX processor, you should see the following message in the command window when you start a simulation: "No physics hardware present, using software physics." This message will go away after you install the processor. You will also have access to additional properties in the AGEIA properties dialog box. You can access this dialog box through Start, Control Panel, and AGEIA PhysX. (If you are using Windows Vista, switch Control Panel to Classic View first, and then double-click AGEIA PhysX.) The Settings tab allows you to reset the AGEIA PhysX card and to start extended diagnostics. These functions will be available only if you have installed the card on your development machine.

Tracking the Frames per Second

VSE allows you to track the frames rendered per second using an optional status bar. The frames per second (FPS) setting is highly dependent on the capabilities of the graphics card and system capabilities of the development machine. It is also dependent on the number of entities and complexity of the objects within the simulation. The exact number of objects that you can add to a simulation depends on things such as the number of polygons used in those objects. A simple shape entity may have only a handful of polygons, but a more complex entity such as a robot may have thousands.

When using VSE to render a simulation, the FPS setting will be limited to the refresh rate for your monitor. Even if you were to upgrade your graphics card and install more memory, your simulation would not render faster than approximately 58 frames per second. The only exception to this would be if you disabled rendering and produced no visible image.

> **Note** You can see what refresh rate your monitor is using by opening the Display Settings dialog box and then clicking Advanced Settings. The Monitor tab lists the screen refresh rate, which is typically 59 Hertz for LCD monitors and 60 or more for CRT monitors.

If you were to run the simulation named "Basic Simulation Environment" (which is supplied with MSRS) and look at the FPS, you would see that it is close to 58 FPS. (You can view the FPS by choosing View, Status Bar, and then reviewing the contents of the Status Bar.) Alternatively, if you then were to run the simulation named "Simulation Environment with Terrain," you might see a drop in the FPS. Whether there would be a drop in FPS would depend on the type of graphics card you were using.

The terrain simulation (see Figure 4-3) includes numerous simply shaped objects such as spheres, boxes, and capsules. It also includes more complex objects such as tables, a robot, and even the terrain itself. To get an idea of how these objects are constructed, you can run the simulation and create a wireframe rendering using the Render menu item. The wireframe rendering shows the lines between all vertices but does not fill them in. This displays all the polygon shapes involved in forming the entity.

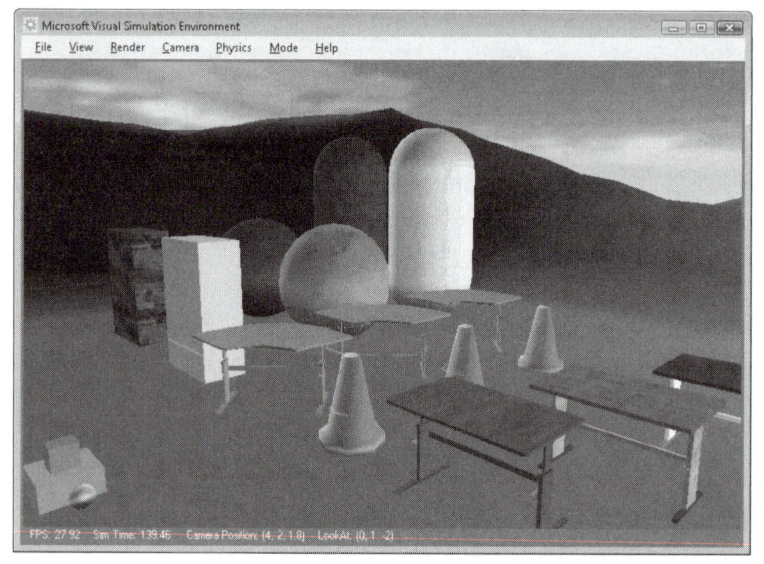

Figure 4-3 The Simulation Environment with Terrain, a simulation provided with MSRS, renders a variety of objects.

The only way to have the simulation run faster is to disable rendering by selecting the Render menu item and then selecting No Rendering. Even though you will not be able to see the rendered results, you could use the Log Messages feature from the simple dashboard service to record messages while controlling the simulated robot.

Running a Simulation

You can run a simulation and initiate VSE by executing DssNode and loading a manifest file that includes the simulation engine service. Recall from Chapter 2 that a manifest file provides a list of service contracts. These service contracts contain information that other services need to work with the target service. A manifest file that loads the simulation service will include a Partner node such as the following:

```
<Partner>        <Contract>http://schemas.microsoft.com/robotics/2006/04/
simulationengine.html
</Contract>
<Service>IRobot.Create.Simulation.xml</Service>
<Name>StateService</Name>
</Partner>
```

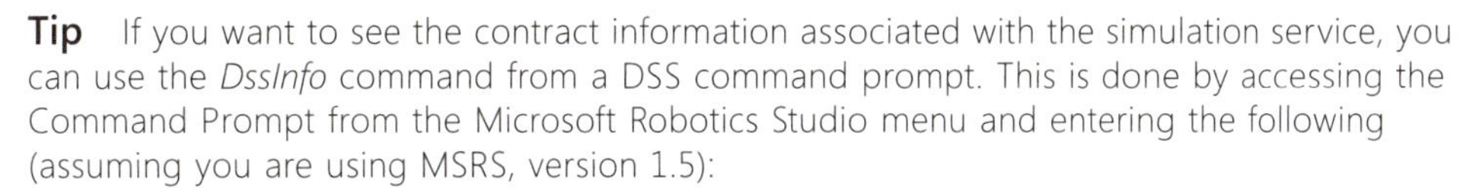

> **Tip** If you want to see the contract information associated with the simulation service, you can use the *DssInfo* command from a DSS command prompt. This is done by accessing the Command Prompt from the Microsoft Robotics Studio menu and entering the following (assuming you are using MSRS, version 1.5):
>
> **DssInfo "C:\Microsoft Robotics Studio (1.5)\bin\simulationengine.proxy.dll"**
>
> This results in the listing of information about the service such as the service contract name and which PortSet operations are supported.
>
> Alternatively, you can select Run Dss Node from the Microsoft Robotics Studio menu, which starts a DSS node and loads the System Services in an Internet browser window. You can then click Control Panel in the left pane and scroll to the Simulation Engine service. Clicking the link for that service displays the information in an Internet browser.

MSRS includes several simulations that you can find on the Microsoft Robotics Studio menu and in the Visual Simulation Environment folder. These simulations represent a variety of scenarios that demonstrate how to work with VSE and build simulations using the entity objects supplied with MSRS. The simulations included with MSRS 1.5 include the following:

- **Basic simulation environment** The basic simulation includes only a few entities, which include a large sphere that looks like the world and a small grey box. No robots are included with this simulation, which is meant to teach the fundamentals of working with VSE. Because the scene contains a few basic objects and includes sky and ground entities, it can be a useful starting point for readers wanting to create their own simulations. Readers can locate the source code for the simulation (which is Simulation Tutorial 1) in the \samples\SimulationTutorials\Tutorial1 subdirectory.
- **IRobot Create simulation** This simulation includes only one robot—the iRobot Create. The command line for this simulation loads two manifest files: iRobot.Create.Simulation.Manifest.xml and SimpleDashboard.Manifest.xml. The Simple Dashboard is a service included with MSRS that allows you to control robots using the services available with the simulation. Readers can locate the source code for the Simple Dashboard in the \samples\misc\SimpleDashboard subdirectory.
- **KUKA LBR3 Arm simulation** This simulation (which is Simulation Tutorial 4) features the KUKA LBR3 articulated arm and several large boxes shaped like dominos. This six-axis robotic arm allows you to move the arm by using the Simple Dashboard to adjust the angles of each joint. The source code that accompanies the KUKA LBR3 Arm simulation included with MSRS is found in the \samples\SimulationTutorial4 subdirectory.

> **Tip** KUKA provides access to additional tutorials that feature its robotic arms. Refer to the following URL for a listing of available tutorials, along with the source code: *http://www.kuka.com/usa/en/products/software/educational_framework/*. Readers interested in working with the KUKA arm should review the tutorials found on the KUKA Web site because they demonstrate the wide variety of capabilities for the multi-jointed robot.

- **LEGO NXT Tribot simulation** This simulation features the LEGO NXT Tribot and several simple objects such as spheres, capsules, and a yellow cone. Even though the source code for the simulation should be located in the \samples\SimulationTutorials \Tutorial6 subdirectory, this was not included with the MSRS 1.5 installation. You can access this tutorial by going to *http://msdn2.microsoft.com/en-us/library/bb483043.aspx*.
- **Multiple simulated robots** This simulation (which is Simulation Tutorial 2) features the Pioneer 3DX and LEGO NXT Tribot robots, along with a table. Using the Simple Dashboard and the drive services associated with each robot, you can drive the robots around the environment and even force them to collide. Readers can locate the source code for the simulation in the \samples\SimulationTutorials\Tutorial2 subdirectory.
- **Pioneer 3DX simulation** This simulation features the Pioneer 3DX and several basic entities such as boxes, capsules, and spheres. These additional entities act as obstacles, and you can use this simulation to see the effects of using the laser rangefinder device that comes with the Pioneer 3DX. While the simulation is running, obstacles within the laser rangefinder's field of view will flash red, dotted lights across the surface of the obstacle. You can use the Simple Dashboard service to view data returned from the robot as it is being moved around the simulation scene.
- **Simulation environment with terrain** Rather than operating the robots on a flat surface as is done in the other simulations, this simulation features a hilly terrain. The simulation features a Pioneer 3DX robot, but it does not look like the same robot featured in other simulations because the mesh object file associated with the entity has been removed. The robot will still perform the same, but it will not look as realistic without the mesh file. Readers can locate the source code for this simulation in the \samples\SimulationTutorials\Tutorial5 subdirectory.

> **Note** Even though the documentation that accompanies MSRS lists six simulation tutorials, the \samples\SimulationTutorials directory contains directories for tutorials 1, 2, 4, and 5. There are also discrepancies between the simulations included in the MSRS menu item, the code available with the installation, and the help file included with MSRS.

When running a simulation, the first thing you notice is that an instance of the command window appears. The simulation uses the command window to start a DssNode and also to load one or more manifest files. This loads the simulation engine service and initiates VSE. By default, VSE starts the simulation in run mode and renders the scene as a solid 3D image. Alternatively, you could render the scene in a wireframe, physics, or a combined view. Because rendering of entities can be an expensive process when the scene involves a large number of entities, it is possible to turn rendering off.

Tip If you encounter errors when trying to run any of the simulations provided with MSRS, open the solution file in Visual Studio and rebuild the solution. MSRS uses strong name signing, and it is possible for some services to be recompiled and for this to cause problems with other projects.

To see which files are associated with a simulation, click on the Start menu, All Programs, Microsoft Robotics Studio (1.5), and then Visual Simulation Environment. Right-click the icon for the simulation, and then choose Properties. The target should list all manifest files that are loaded for this simulation. There might be more than one manifest loaded, so you might have to recompile more than one solution. For example, the iRobot Create simulation loads two manifest files: IRobot.Create.Simulation.Manifest.xml and SimpleDashboard.manifest.xml.

Note that it might take a long time for some of the solution files to rebuild because they might contain several projects. Make sure the Visual Studio status bar states that the rebuild all was successful before attempting to run the simulation again.

Using the Simple Dashboard Service

MSRS provides a service that allows you to operate a variety of robotics hardware. The Simple Dashboard service initiates a Windows-based dialog box (see Figure 4-4) that allows you to connect to a DSS node and initiate one or more services running on that node. The dashboard appears automatically whenever you run a simulation that partners with the Simple Dashboard service. Most of the simulations provided with MSRS include the dashboard.

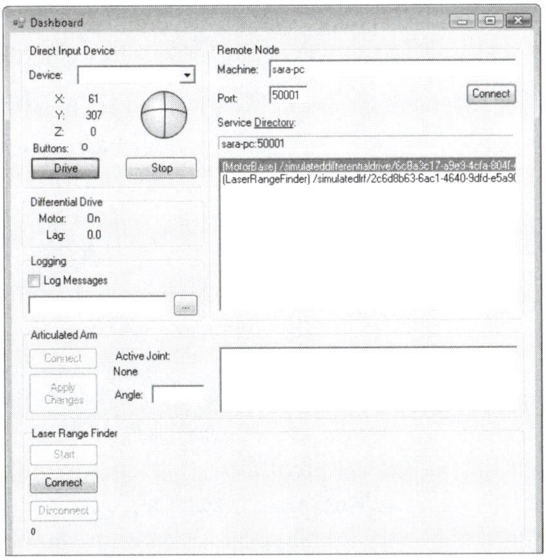

Figure 4-4 The Simple Dashboard service is one of the miscellaneous services offered with MSRS. This service initiates a Windows-based dialog box that allows you to control your physical or simulated robot.

> **Tip** When you run a simulation that includes the dashboard, the dashboard form will not receive focus and might be lost behind other windows running on your desktop. You might have to move the VSE window before you can see the dashboard form.

To connect to a DSS node on your local development machine, enter **localhost** in the Machine text box for the Dashboard dialog box. You then click Connect, and a list of services running on that node should appear in the list box below Service Directory. These services would have been loaded by the manifest file when you initiated the simulation. Depending on the service selected, you can perform the following functions:

- **Drive the robot** For robots with a differential drive system, such as the iRobot Create and Pioneer 3DX, a simulated differential drive service can be used to drive the robot. To do this, you must first select the drive service from the list of available services. You must then click the Drive button on the left side of the dialog box. At this point, you can control the robot using the mouse, trackball, touchpad, or joystick.

> **Tip** It is possible to use an Xbox 360 controller to drive robots in a simulation. Readers interested in doing this should refer to the documentation available with MSRS concerning the XInput controller service.

- **Move an articulated arm** For robots that feature an articulated arm, such as the KUKA LBR3, you can change the angles for each active joint to move the arm within the simulation. To control the arm using the Simple Dashboard dialog box and the KUKA LBR3 simulation, select the simulated LBR3 arm service from the list of available services. At this point, the Articulated Arm section should list all the joints for the simulated arm.
- **Monitor data from a laser rangefinder device** For robots that include a laser rangefinder, such as the Pioneer 3DX, you can view the results obtained from the device. The results are used to indicate whether the robot encounters an object and measures the distance from the robot to the obstacle.
- **Log messages** The Simple Dashboard allows you to log messages as the simulation is executing. To do this, you need to specify a path to a log file, and you must store the log file beneath the Store subdirectory, which is part of the MSRS installation. The resulting XML-based file will include state that is retrieved by performing DSS operations. Data will continue to be written to the log file as long as the simulation is running. Recording data to the log file allows you to go back and see precise values returned from robot sensors. For example, you can log messages while operating the laser rangefinder and then go back later to see the exact distance measurements obtained while the simulation was running.

Note Although you can view messages as the simulation is running, the file size for a newly created log file remains at zero bytes until the simulation ends.

Enhanced Version of the Simple Dashboard Service

Because MSRS includes code for services such as the Simple Dashboard service, it is possible for you to extend the functionality available in these services. Ben Axelrod (see the following profile sidebar), a developer for Coroware (*http://www.coroware.com*) has released an enhanced version of the Simple Dashboard service (see Figure 4-5) that he makes available as a free download on his Web site (*http://www.benaxelrod.com/MSRS/index.html*). Be sure to put "MSRS" in all capital letters as this URL is case-sensitive.

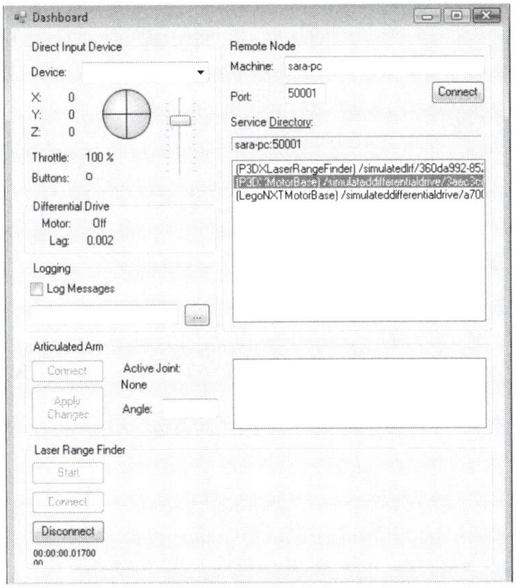

Figure 4-5 An enhanced version of the Simple Dashboard service, available on Ben Axelrod's Web site.

Profile: Ben Axelrod

For Ben Axelrod (see Figure 4-6), the drive to build robots began in early childhood, but it was not until high school, when his grandparents bought him a LEGO Mindstorms system, that his passion blossomed. At the age of 27, Axelrod has accomplished quite a lot in a field he almost missed out on.

Figure 4-6 Ben Axelrod, robotics developer for Coroware.

Two-time winner of the LEGO Mindstorms Novice Hall of Fame (*http://mindstorms.lego.com/eng/inventions/default.asp*), Axelrod has already held positions at Microsoft, Iguana Robotics Inc., and Coroware. In 2003 he graduated summa cum laude from Syracuse University in mechanical engineering and went on to earn a master's degree in computer science from the Georgia Institute of Technology.

But it was Axelrod's work with the LEGO Mindstorms that allowed him to see the world through the "eyes" of a robot. His inventions not only earned him the LEGO Hall of Fame distinction, but awakened a creative energy that has allowed him to work toward inspiring others to become roboticists.

When asked to speak about robots to a group of kids in Illinois, Axelrod went above and beyond the call of duty and produced a video that gives a realistic overview of the current world of robotics. The 10-minute video is available for download from Axelrod's Web site through the following URL: *http://www.benaxelrod.com/robotvideo/index.html*. When asked why he likes talking to kids about robotics, Axelrod replied:

> *I almost missed the chance to be an engineer and roboticist. I just didn't know that this was a possible career choice, and I want to show kids that this is possible. I also feel that engineering and computer science are very important jobs, and robotics is a great way to interest kids.*

The enhanced Simple Dashboard service provides a few time-saving features, such as defaulting the localhost value for the machine name and removing the Drive and Stop buttons. The service also features many improvements to the Direct Input Device, which is used to drive robots with a differential drive system. The original dashboard service was limited to 60 percent power when using the joystick to drive the robot. The enhanced version allows you to drive the robot at 100 percent power and to control this using a slider control added to the dashboard form. In addition to controlling the robot with the mouse or joystick, you can use arrow keys on the keyboard.

Enhancements were also made to the laser rangefinder section, and the extended service features two views: a cylinder view and an overhead view. Green dots project the robot width forward, and red colored obstacles indicate objects directly in front of the robot.

Readers can use the enhanced dashboard by first downloading the zipped file from the following URL: *http://www.benaxelrod.com/MSRS/index.html*. Because the solution includes references to several projects that are available with MSRS, you should extract and store the contents of the downloaded file in the \samples\misc folder. The files available in the extracted folder should replace the files in the SimpleDashboard folder, which is already in the misc subfolder. Before replacing the contents of the SimpleDashboard folder, copy the original files to another folder.

After you copy the files to the correct location, you just need to open the solution file using Visual Studio and rebuild the solution. From this point on, whenever simulations involving the Simple Dashboard service are loaded, the enhanced version of the dashboard should appear.

Note Because of strong name binding, you may have to rebuild some of the dependency projects, such as the SickLRF project. To do this, you will need to open the dependency project in Visual Studio and do a rebuild of this project before returning the SimpleDashboard solution file.

Editor Settings

VSE allows you to modify graphics settings that will apply to multiple simulations. Although you can set additional settings using code in a DSS service, the graphics settings (see Figure 4-7) apply to commonly used settings such as the brightness or quality level of the scene. The graphics settings, which include the following, are accessible through the Render menu item:

- **Exposure** This affects the brightness of the simulation scene. By default, the exposure is set with a value of zero. Setting the value higher makes the scene lighter, and setting it to a negative value makes the scene darker.

- **Antialiasing** Depending on the capabilities of your graphics card, you may not have any options for this setting. Alternatively, you could have multiple settings that control the level of antialiasing applied. Antialiasing controls how smoothly edges are rendered and, thus, can affect the quality of your simulation scene. Without any antialiasing, edges can appear jagged and not smooth. As long as the graphics card performs the antialiasing on the hardware itself, setting this to a higher value should have little or no impact on the rendering speed.
- **Rotation Movement Scale** This controls how sensitive the mouse or touchpad is when moving around a scene. By default, this is set to a value of one. Setting it higher will make the mouse move faster, and setting it to a negative value makes the mouse move slower.
- **Translation Movement Scale** Similar to the Rotation Movement Scale setting, this controls how sensitive the keyboard and joystick are when moving around a simulation scene. Setting it to a higher value makes the controls work faster, and setting it to a negative value makes them move slower.
- **Quality Level** This controls the overall quality of the rendered scene and can be set to Low, Medium, or High. By default, VSE will select the level most appropriate for your graphics card. Setting the value higher than the value recommended by VSE will likely affect the rendering speed of your simulation.

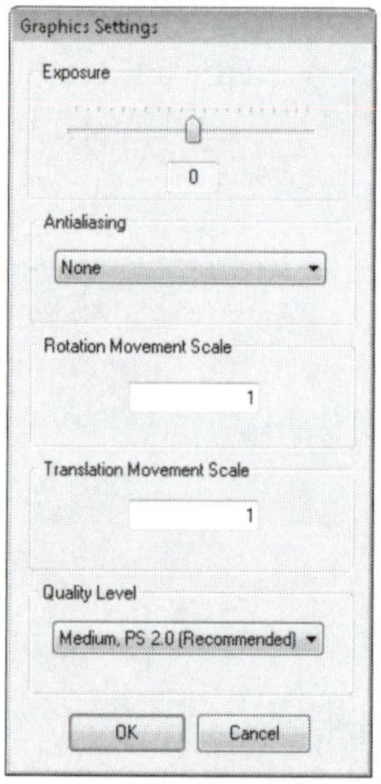

Figure 4-7 The Graphics Settings dialog box provides access to commonly used settings that apply to multiple simulations.

You can also modify physics settings through the Physics menu item. The Physics settings allow you to control how the physics engine applies principles such as gravity. The gravity setting, which defaults to a value of 9.81, controls how the physics engine will handle gravity within the simulation scene. Typically, you will want to keep this setting as the default, but you may want to turn it off by setting it to a value of zero while making changes to the entities in edit mode.

Working with Coordinates

Unless you have spent the last several years working intimately with vector maths and computer simulations, you may feel a bit overwhelmed by some of the terms and concepts used by the simulation engine. Do not worry about it too much. MSRS hides much of the complexity from you and acts as a wrapper for functions available in the Microsoft DirectX SDK, Microsoft XNA Framework, and AGEIA PhysX SDK. This section will give you a high-level overview of the things you need to understand when working with objects in 3D space.

To represent a physical object in 3D space, you must work with coordinates that represent the object's position. These coordinates consist of points along the X, Y, and Z axis lines. The MSRS simulation engine uses what is known as a right-handed coordinate system. This affects which direction the positive Z-axis points in. Figure 4-8 is a diagram that indicates which direction the X, Y, and Z axis lines move toward. Lines moving in the opposite direction indicate points along the negative axis.

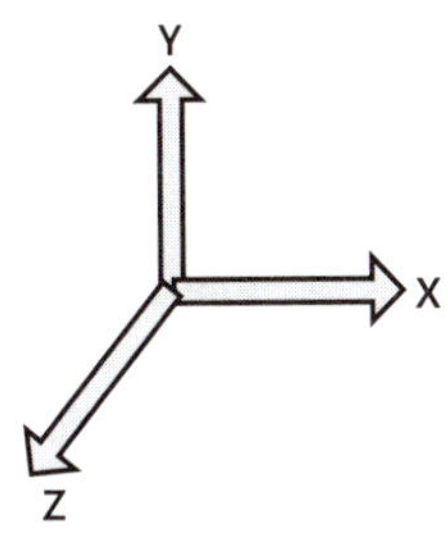

Figure 4-8 The simulation engine uses a right-handed coordinate system, which indicates the direction the positive Z-axis points toward.

When creating or editing entities, you will be able to edit the coordinates to change things such as the entity's position. By changing the Y-axis coordinate to a higher positive value, the entity will move farther up in the simulation scene. If the value is set to a lower value, the entity moves farther down.

> **Tip** For readers interested in running a robot through a simulated maze, you can refer to a maze simulation tutorial hosted on Trevor Taylor's Web site (*http://sky.fit.qut.edu.au/~taylort2/MSRS/MazeSimulator/MazeSimulator.htm*). Taylor is a faculty lecturer in the Information Technology department at the Queensland University of Technology. He is also an enthusiastic roboticist and MSRS expert, especially in the area of simulation. Check out his MSRS Web page at the following URL for more interesting information: *http://sky.fit.qut.edu.au/~taylort2/MSRS/*.

What is an Entity?

An entity represents a physical object that is used within a simulation. This can include, not only robots, but obstacles within the environment such as tables and chairs. It can also include ground and sky objects, and simulations can take place either in outdoor or indoor environments.

An entity can be hierarchical and involve parent and child relationships. This is the case when dealing with most robotics hardware where the motor base object represents the parent entity. Child entities would be the robot's sensors and actuators. For example, the Pioneer 3DX simulated robot entity has three child entities: the bumpers, a laser rangefinder, and a Web camera.

Even the most basic simulations may contain a dozen or more entities. Complex simulations could involve hundreds or even thousands of entities. All simulations require a ground entity–unless you happen to be simulating robots floating in space. They will also require a sky object, but they do not require entities such as the sun or other elements that might appear in the sky. If the simulation is done indoors, it will typically be rendered inside a box, and you generally do not need to include things like light fixtures.

If you want to see the effect of numerous objects on a simulation, you can use one of the simulation tutorials and a Web browser to quickly add table objects to the simulation. To do this, you need to run Simulation Tutorial 2, which is available in the Visual Simulation Environment folder as the Multiple Simulated Robots simulation. When you first run this simulation, a table and two robots will appear in the simulation scene. While the simulation is running, open a Web browser and enter the following URL: ***http://localhost:50000/simulationtutorial2***.

Your Web browser should display the state for the simulation in the form of XML. The state should contain a variable that indicates the number of tables in the simulation. If you continue to hit the refresh button in your Web browser, this number will increase by one, and tables will fall down from the sky onto the simulation scene (see Figure 4-9).

As additional entities are added to a simulation, it can cause the simulation to slow down and render at lower FPS. For this reason, you should try to limit the number of entities included in a simulation. MSRS includes several entity types (refer to Table 4-2 for a listing of these entity types) that you can use to build most of the entities you will need. The exception to this is robotics hardware not supported by MSRS and complex obstacles.

Tip By default, the status bar is not displayed in VSE. To see the FPS settings, you will need to click on the View menu and then select Status Bar. The FPS setting will continually change even if nothing is changing within the simulation scene. This is due to things such as the refresh rate for your monitor and available CPU resources. The status bar also displays the simulation time, camera position, and Look At position.

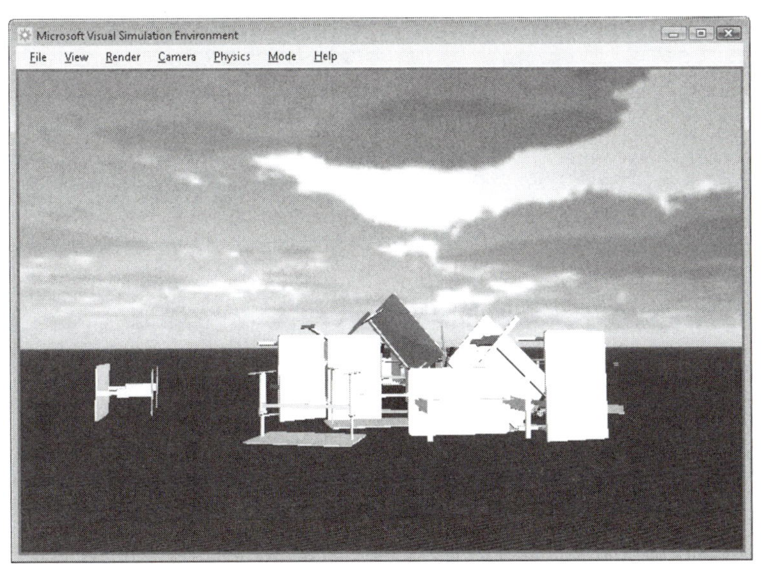

Figure 4-9 Effect of adding multiple table objects to Simulation Tutorial 2 using a Web browser.

Create a New Entity

To create a new entity in VSE, you must be in edit mode within the simulation environment. While in edit mode, the top-left pane will display all of the entities that currently appear in the simulation scene. When you select an item, the properties for that entity appear in the bottom-left pane. Table 4-1 lists the basic properties available for an entity.

Depending on the type of entity, there may be additional properties available. For example, an entity representing the Pioneer 3DX robot will include a property named *MotorTorqueScaling*, which is used to set a scaling factor that applies to all motor torque requests. Obviously, this is a property that would have no use for a sky entity. A sky entity does, however, have use for properties related to fog, such as the fog color.

Table 4-1 Basic Entity Properties

Property Name	Description
EntityState	Associated with a set of additional properties that control how the entity appears and behaves. Refer to Table 4-2 for a listing of the properties associated with entity state.
ServiceContract	Used to specify the contract for a service. The contract is associated with an HTML file, and you would need to include the entire path to the contract. When an entity is associated with a service contract, it will be included in the manifest file that is generated when a scene is saved.
InitError	Contains a description of any errors that occurred during the initialization of the entity.

Table 4-1 Basic Entity Properties

Property Name	Description
Flags	Controls how the entity is rendered and can be set with one of the following options: ■ **None** No rendering properties set. This is the default setting. ■ **UsesAlphaBlending** Allows for transparency and, thus, more realistic scenes. Depending on the capabilities of the graphics card, using alpha blending can degrade performance. ■ **DisableRendering** Turns off rendering. ■ **InitializedWithState** Indicates that the entity was built by modifying the XML associated with the simulation scene. ■ **DoCompletePhysicsShapeUpdate** Forces an update of all physics shapes that are part of the entity. ■ **Ground** Used to specify that this is a ground entity.
ParentJoint	Specifies the joint used to connect the entity to the parent entity. This is used when you are dealing with entities composed of other entities, such as a robot with legs and/or arms.
Position	Specifies the X, Y, and Z coordinates that control where the entity is located in the simulation scene. The coordinates are specified in meters.
Rotation	Based on Euler angles, this property specifies the radians or degrees that control the orientation of the entity.
Meshes	Specifies one or more meshes associated with the entity. Meshes will be added with the VisualEntityMesh collection editor. Materials are used to form physical objects, and entities can be associated with one or more materials. It is also possible for an entity to have no materials, but this will mean the entity is not visible.
MeshRotation	Specifies the yaw, pitch, and roll values (which are stored in X, Y, and Z coordinate points), which control how the mesh object is rotated. Typically, these will be set initially with 0 values.
MeshTranslation	Specifies the X, Y, and Z coordinates that control how the mesh object is translated, relative to the origin of the entity. Typically, these will be set initially with 0 values.
MeshScale	Specifies height, width, and length values (which are stored in X, Y, and Z coordinate points), which control how the mesh is scaled. Typically, you will initially set the coordinates with a value of 1. Higher values make the object appear larger, and negative values make the object appear smaller.
Shapes	The availability of this property depends on the type of entity. For some entities, this property will not be available at all. When the entity is based on a single shape, such as a cylinder, capsule, or box, you can specify what type of shape and set additional properties associated with that shape in a separate dialog box. If the entity is based on a specialized type that can have more than one member, such as a bumper array or wheel, then a collection editor dialog box is used to specify properties associated with each member of the array.

While in edit mode, hold down the Ctrl key to see which entity is selected. The selected entity will appear with a circle surrounding it, as shown previously in Figure 4-1. You can then use the arrow keys to move the main camera around the object as it is zoomed in. This allows you to see all sides of the object clearly. For example, Figure 4-10 shows a close-up of the rear view for the LEGO NXT. You can also use the Ctrl key and mouse to change the entity's position or rotation. This is done by selecting the object in edit mode and also selecting the position or rotation property. While you hold down the Ctrl key, use the mouse to move or rotate the entity.

Figure 4-10 Using the Ctrl and arrow keys, you can focus in on a selected entity and view all sides such as the rear view of the LEGO NXT.

To create a new entity, click on the Entity menu, and then select New. This brings up the New Entity dialog box (see Figure 4-11) and allows you to define a new entity. The entity can be added to either the current simulation or to another assembly. An entity type (see Table 4-2 for a list of available entity types) must be selected. The entity type acts as a template for an entity and, thus, assigns initial property values for a new entity.

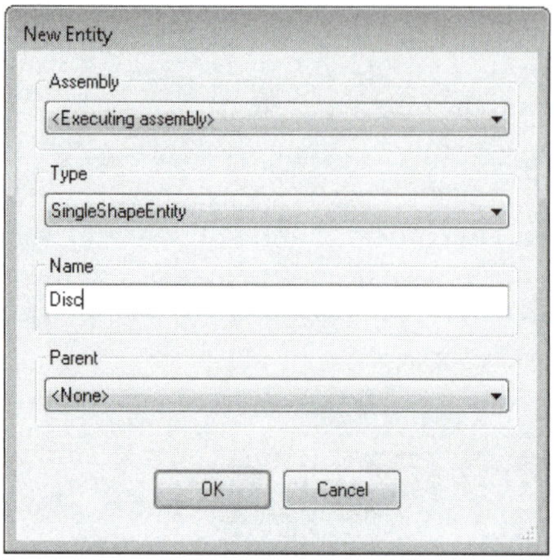

Figure 4-11 While in edit mode, the New Entity dialog box is used to add new entities to a simulation scene.

Table 4-2 Entity Types Available in MSRS

Entity Type	Description
ArmLinkEntity	Used to model one link of an articulated robotic arm, such as the KUKA LBR3, which has six degrees of freedom. As opposed to the KukaLBR3Entity type, this type represents just one portion of the robotic arm, and the entire arm would consist of seven links.
BumperArrayEntity	Used to create an array of bumper sensors. The entity created from this type will act as a child for a robot that uses bumper contact sensors (such as the Pioneer 3DX and iRobot Create robots).
CameraEntity	Used to create a camera entity that has a 90-degree field of view. This affects how much of the simulation scene appears in a single frame. This entity will act as a child entity for a robot that hosts the Web camera (such as the Pioneer 3DX or iRobot Create robots).
HeightFieldEntity	Used to represent a ground object and can be used to represent a flat ground. By passing in an array of height samples, you can represent the surface or an uneven terrain.
iRobotCreate	Used to model an iRobot Create robot.
KukaLBR3Entity	Used to model a KUKA LBR3 robotic arm.
LaserRangeFinderEntity	Used to create a laser rangefinder entity, which will act as a child for a robot that uses a laser rangefinder (such as the Pioneer 3DX).
LegoNXTTribot	Used to model a LEGO NXT Tribot.
LightSourceEntity	Used to create light-based entities, such as the sky or a light fixture. Light sources can be either directional, in which the light rays come from one direction, or omni, in which the light comes from all directions (like the flame from a candle does).

Table 4-2 Entity Types Available in MSRS

Entity Type	Description
MultiShapeEntity	Used to define entities that consist of multiple entities. For example, the table entity that is represented in several of the simulations is built using a MultiShapeEntity type. Each shape in a MultiShapeEntity has a fixed pose in relation to the other entities.
Pioneer3DX	Used to model a MobileRobots Pioneer 3DX robot.
SimplifiedConvexMesh-EnvironmentEntity	Used to generate a convex hull mesh, which specifies the surrounding area for the entity. Even though the entities may appear the same visually because of the mesh object, the entity can appear different while rendered as a physics model. This will affect how collisions with other objects are handled. See the documentation for Simulation Tutorial 5 for an example of how to use this entity type when creating table- and cone-shaped entities.
SingleShapeEntity	Used to create an entity with a single shape, such as a sphere, box, or capsule.
SkyDomeEntity	Used to render a sky using a dome two-dimensional (2D) image, rather than the cube map, which is used to create a SkyEntity. Using a 2D image is less resource-intensive and should be used unless a finer level of detail is required.
SkyEntity	Used to render a sky using a cube map, which is a six-sided image file. Two cube maps can be used: one to draw the sky and the other to light the scene. See the Simulation Tutorial 5 documentation for more information about using the SkyEntity entity type. Because the SkyEntity type involves cube maps as opposed to a 2D image, it will consume more processing power to render. If possible, the SkyDomeEntity type should be used to render a sky entity.
TerrainEntity	Used to create a complex ground entity that uses a bitmap image.
TerrainEntityLOD	Also used to create a complex ground entity, this entity type renders a finer level of detail. See the documentation for Simulation Tutorial 5 for an example of using this and the TerrainEntity entity types.
TriangleMeshEnviron-mentEntity	This entity type is similar to the SimplifiedConvexMeshEnvironment-Entity except that the entity will be concave in shape. This means that the physics engine will have to work harder when working with this entity, so, generally, you will only want to use this entity type when absolutely necessary. See the documentation for Simulation Tutorial 5 for an example of how to use this entity type.
WheelEntity	This represents an actuator and will be the child entity for a robotic motor base.

Table 4-3 lists the basic properties available when you select the ellipses next to the *EntityState* property. Depending on what type of entity it is, the properties that appear might be slightly different. These properties control both how the entity appears when rendered and how it behaves in relation to the physics engine.

When setting the properties listed in Table 4-3, consider that MSRS represents distance in terms of meters and weight in terms of kilograms. In case you do not remember, 1 meter equals 1.093 yards or 39.37 inches. Alternatively, 1 yard equals .0914 meters or 36 inches. If you want to compute the distance of an entity in meters, you could multiply the number of yards by .0914. If you only knew the number of inches, you could multiply the number of inches by .0254. Both calculations should give you a value in meters.

Also, 1 kilogram equals 1,000 grams, and 453.59 grams equals 16 ounces or 1 pound. If you want to compute the weight in kilograms of an object weighing 3 pounds, you would first need to multiply the number of pounds by 453.59 (3 * 453.59) to arrive at a value of 1,360.77. To get the number of kilograms, divide that number by 1,000 (1360.77 / 1,000) to arrive at a value of 1.36. This would be the value you would set for the approximate number of kilograms needed to represent 3 pounds.

Table 4-3 Basic Entity State Properties

Property Name	Description
DefaultTexture	Refers to a texture file that applies to all shapes for that entity. The texture file should be a DirectX texture file, which will have a .dds file extension. The texture file applied through the entity state is optional and is only needed if a mesh object is not specified.
Effect	Refers to a special graphics effect file (FX) that can be used to render the entity. The graphics effect file is a script that tells DirectX what shaders to load. MSRS comes with several effect files (each with an .fx file extension), which can be used to accomplish various rendering effects for entities such as the sky and the terrain. The FX files supplied with MSRS can be found in the \store\media subdirectory.
Mesh	Refers to an optional mesh file that can be used to render the entity. A mesh object file, which has an .obj file extension, is created with a 3D graphical editing tool capable of exporting to an alias object format. The mesh is what makes the entity appear realistic, and, without it, the entity would appear as just a basic shape(s). You can also specify an optimized binary file, which allows the entity to render more quickly. The binary file has a .bos file extension. MSRS provides a command-line converter tool named Obj2bos.exe that you can use to convert the mesh files to a binary format. The mesh files available with the MSRS installation are located in the \store\media folder.

Table 4-3 **Basic Entity State Properties**

Property Name	Description
Flags	Specifies a flag, which affects how the entity is rendered and how it behaves. It can be set with one of the following options: ■ **Kinematic** Indicates that the object has infinite mass and, therefore, will remain unmovable ■ **IgnoreGravity** Allows you to disable the effects of gravity for a single entity ■ **DisableRotationX** Prevents the entity from rotating along the X-axis ■ **DisableRotationY** Prevents the entity from rotating along the Y-axis ■ **DisableRotationZ** Prevents the entity from rotating along the Z-axis
Name	This is a unique name associated with the entity.
AngularVelocity	Used to specify the speed and direction at which an entity is rotating. This is specified through X, Y, and Z coordinates points.
Velocity	Used to set the X, Y, and Z coordinates associated with the entity's linear velocity.
AngularDamping	Used to set an integer coefficient that is associated with how an entity rotates through space in relation to its mass. The default value of 0 indicates that no damping will take place.
Density	Used to specify the density of an object, which represents the mass of an object divided by its volume. This setting is calculated by the physics engine, so there is no need to set the value to anything other than the default value of 0.
InertiaTensor	Used to specify the X, Y, and Z coordinates that control how mass is distributed across an entity. This setting is calculated by the physics engine, so there is no need to set the value to anything other than the default value of 0 for the X, Y, and Z coordinates.
LinearDamping	Used to set an integer coefficient associated with the friction produced by an entity as it moves through space. The default value of 0 indicates that no damping will take place.
Mass	Used to specify the mass of an entity in kilograms. The value assigned to the mass property is important when considering simulations that involve potential collisions. Because mass and gravity affect how much an object weighs, the higher the value for mass, the more the object will weigh. What occurs when a robot collides with another robot depends a lot on how heavy both the robots are.

The values associated with the state of an entity will change as the simulation is running. For example, the robot's position will be determined by the X, Y, and Z coordinate values set for the position property. The initial values set for this property will change as the robot is driven around the simulation scene. You can see this for yourself by running a simulation, such as the iRobot Create simulation, and switching into edit mode. Initially, the position of the Create robot is set with the following coordinate values: 2, 0, 0. If you switch back to run mode and use the Simple Dashboard to move the Create robot, you would find that the position values have changed when you return to edit mode.

Disabling and Enabling Physics Mode

When you select edit mode, VSE will automatically turn off the physics engine. Even though you can turn it back on by clicking Enable from the Physics menu, it is generally not advisable to do so. When physics is enabled, certain things are not possible, such as two objects occupying the same point in space at the same time. This type of conflict can occur when you copy and paste an entity because the pasted version of the entity will have the same position coordinates as the original.

If you use copy and paste to create a duplicate copy of an entity while in edit mode, the new entity will appear in the top-left list of entities but will not be immediately visible in the simulation. This is because the copied entity is located in the exact same position as the original, and, thus, only one is visible. If you were to turn physics back on, you would notice that the two entities quickly move apart as if they had suddenly been thrown together and were bouncing off each other.

Alternatively, if you were to try and perform the same copy and paste operation while physics mode is enabled, then the additional entity would immediately be thrown away from the original entity. The choice of whether to enable or disable physics while in edit mode is largely a matter of personal preference and depends on what you are trying to accomplish.

> **More Info** More information or additional information about how to work with the menu items in VSE, refer to the documentation for Simulation Tutorial 6 in the MSRS help file.

Editing the Simulator State File

VSE allows you to save simulation scenes by clicking on File and then selecting Save Scene As. This will create two XML-based files: one is a manifest file and will contain a list of service contacts that need to be loaded to run the simulation. The other file contains information about the state of the simulator and the entities that exist within the simulation scene.

The simulator state file is a snapshot of what the simulation looked like at an exact point in time. The information within this file can be useful for many purposes. While learning about simulation and the numerous properties available for each entity, it can be helpful to browse through the simulator state file and see what values are associated with each property. For example, the following XML is only a portion of the state associated with a single-shape entity named greybox:

```
<BoxShape>
 <phys:BoxState>
   <phys:Name>greybox</phys:Name>
   <phys:ShapeId>Box</phys:ShapeId>
   <phys:Dimensions>
     <physm:X>1</physm:X>
     <physm:Y>1.5</physm:Y>
     <physm:Z>0.5</physm:Z>
   </phys:Dimensions>
   <phys:Radius>0</phys:Radius>
   <phys:Material>
     <phys:Name>gbox</phys:Name>
     <phys:Restitution>0.5</phys:Restitution>
     <phys:DynamicFriction>0.4</phys:DynamicFriction>
     <phys:StaticFriction>0.5</phys:StaticFriction>
     <phys:MaterialIndex>2</phys:MaterialIndex>
   </phys:Material>
   <phys:MassDensity>
     <phys:Mass>10</phys:Mass>
     <phys:InertiaTensor>
       <physm:X>0</physm:X>
       <physm:Y>0</physm:Y>
       <physm:Z>0</physm:Z>
     </phys:InertiaTensor>
     <phys:CenterOfMass>
   <physm:Position>
         <physm:X>0</physm:X>
         <physm:Y>0</physm:Y>
         <physm:Z>0</physm:Z>
       </physm:Position>
       <physm:Orientation>
         <physm:X>0</physm:X>
         <physm:Y>0</physm:Y>
         <physm:Z>0</physm:Z>
         <physm:W>0</physm:W>
       </physm:Orientation>
     </phys:CenterOfMass>
     <phys:Density>0</phys:Density>
     <phys:LinearDamping>0</phys:LinearDamping>
     <phys:AngularDamping>0</phys:AngularDamping>
   </phys:MassDensity>
   <phys:LocalPose>
     <physm:Position>
       <physm:X>0</physm:X>
       <physm:Y>0</physm:Y>
       <physm:Z>0</physm:Z>
     </physm:Position>
     <physm:Orientation>
       <physm:X>0</physm:X>
       <physm:Y>0</physm:Y>
```

```
        <physm:Z>0</physm:Z>
        <physm:W>1</physm:W>
      </physm:Orientation>
    </phys:LocalPose>
    <phys:TextureFileName />
    <phys:DiffuseColor>
      <physm:X>0.5</physm:X>
      <physm:Y>0.5</physm:Y>
      <physm:Z>0.5</physm:Z>
      <physm:W>1</physm:W>
    </phys:DiffuseColor>
    <phys:EnableContactNotifications>false</phys:EnableContactNotifications>
  </phys:BoxState>
</BoxShape>
```

In addition to gathering information about the entities used in a simulation, the simulation state file can be useful when there is a need to quickly edit properties or add new entities. After you make changes to a simulation state file, use the File and Open Scene menu items to reload the changes into VSE.

Creating a Simulation from a Service

In addition to creating a simulation scene using VSE, you can create a scene programmatically using Visual Studio. This chapter will not go into great detail about how to do this because the simulation tutorials included with MSRS cover this subject very well. Instead, the chapter will review the basics needed to create a simulation programmatically.

To do this, you will create a new DSS service using the template provided with MSRS. After you create the project, you will need to add references to the following assemblies:

- **PhysicsEngine** Contains methods and properties used to access the underlying AGEIA software physics engine.
- **RoboticsCommon** Contains the *PhysicalModel* namespace, which includes type definitions used to define the physical characteristics of robots. For example, this is where you will find definitions that represent sensors such as analog sensors, contact sensors, and Web cameras. You will also find type definitions for abstract services such as the articulated arm service, which is used to operate a simulated robotic arm.
- **SimulationCommon** Contains the type definitions used when working with both the simulation and physics engines.
- **SimulationEngine** Used to access the simulation engine and perform functions such as keep track of the state for each entity.
- **SimulationEngine.proxy** Proxy for the simulation engine that is used to load the engine as a partner. Proxy assemblies are generated for all compiled services, of which the simulation engine is one.

After you add the assembly references, you will also need to add namespace declarations for the associated references. For example, you need to add the following namespace declarations to the top of the implementation class file for your simulation:

```
using Microsoft.Robotics.Simulation.Physics;
using Microsoft.Robotics.PhysicalModel;
using Microsoft.Robotics.Simulation;
using Microsoft.Robotics.Simulation.Engine;
using engineproxy = Microsoft.Robotics.Simulation.Engine.Proxy;
```

MSRS includes other services (see Table 4-4) that you can use to implement specific functionality. For example, a simulation that includes a robot with bumpers and a differential drive system will need to include references to the SimulatedBumper and SimulatedDifferentialDrive services.

Tip Conscious-Robots.com hosts a Web site specifically dedicated to working with MSRS. On this site, you can download a simulated sonar service (*http://www.conscious-robots.com/en/robotics-studio/robotics-studio-services/simulated-sonar-service-2.html*). This simulated sonar service, which is not included with MSRS, simulates the sonar array for a Pioneer DX3 robot. It uses the Raycast application programming interface (API) in a similar way as the simulated laser rangefinder does.

Table 4-4 **Additional MSRS Services Used to Provide Specific Simulation Functionality**

Service Name	Description
SimulatedBumper	Use to access an array of bumper sensors for a simulated robot. Readers can locate the source code for this service in the \samples\simulation\Sensors\Bumper folder. NOTE: In the MSRS Control Panel, this service is listed as the Simulated Generic Contact Sensors service.
SimulatedDifferentialDrive	Use to drive a simulated robot with a differential drive system. This service uses a reference to the generic differential drive service. Readers can locate the source code for this service in the \samples\simulation\DifferentialDrive folder.
SimulatedLBR3Arm	Use to control the joints for a simulated KUKA LBR3 articulated arm. Readers can locate the source code for this service in the \samples\Simulation\ArticulatedArms\LBR3 folder.
SimulatedLRF	Use to simulate a Sick laser rangefinder device, which is found on the Pioneer robot. This device emits an infrared beam in a 180-degree arc, making it highly accurate at detecting objects. This service uses a reference to the SickLRF service. Readers can locate the source code for this service in the \samples\simulation\Sensors\LaserRangeFinder folder.
SimulatedWebCam	Use to access a Web camera on a simulated robot. Readers can locate the source code for this service in the \samples\simulation\SimulatedWebCam folder.

When setting references to the assemblies listed in Table 4-4, you will need to reference the associated proxy files. For example, the proxy reference used for the Simulated Differential Drive service is named SimulatedDifferentialDrive.2006.M06.Proxy.

> **Note** The solution files included for the simulation tutorials contain the project files for referenced assemblies used by that tutorial. This was done because MSRS uses strong-name signing, and, if you rebuild one dependency, this could affect other projects that use that dependency. By doing a rebuild of the entire solution file, you should be able to avoid the problems that can occur with references not found.

Because the simulation engine is a service, you will need to add it as a partner service. You do this by using the *Partner* attribute, as shown in the following code:

```
[Partner("Engine",
    Contract = engineproxy.Contract.Identifier,
      CreationPolicy = PartnerCreationPolicy.UseExistingOrCreate)]
private engineproxy.SimulationEnginePort _engineServicePort =
      new engineproxy.SimulationEnginePort();
```

In this case, the simulation service was created using the *UseExistingOrCreate* attribute. This means that if the simulation engine is already running, then it should be used instead of trying to instantiate a new instance.

Creating New Entities

The steps required for creating a new entity depend on what type of entity it is. MSRS supports several entity types that can be used to create simulated robotics hardware. It also provides entity types for creating environment entities such as the sky and ground. Both simple and complex objects that represent obstacles can be built using the SimpleShape and MultipleShape entities.

Create Robot Entities

Robot entities are complex and will typically contain one or more child entities. The child entities may represent sensors used on the simulated robot. For example, the simulated version of the Pioneer 3DX robot will have child entities for the laser rangefinder device and bumper sensors. To create the simulated Pioneer entity, you will first need to create the parent entity, which in this case will be the motor base. This can be done using the *Pioneer3DX* class that is included with the simulation engine. This class inherits from the DifferentialDriveEntity, which is used to specify the different physical properties associated with a differential drive system. The following code can be used to instantiate a new entity that represents the Pioneer robot:

```
Pioneer3DX robotBaseEntity = new Pioneer3DX(new Vector3(0, 1, 0));
```

In the previous code snippet, a Vector3 structure was used to hold three coordinate points that represent the robot's position in the simulation scene. MSRS provides vector structures that can be used to store values associated with each coordinate point. There is a Vector2 structure, which stores X and Y coordinate values. Vector3 stores X, Y, and Z coordinates. Finally, the Vector4 structure is used to store X, Y, Z, and W values. When creating the Pioneer robot entity, you will need to assign a unique name for the robot and specify the robot's color using a Vector4 structure.

After you create the motor base entity, you can start the service that represents the Pioneers motor service. To do this, you will need to use the *CreateService* method to create the service as an entity partner. This tells the simulation service what other services should be bound to it at runtime. The following code can be used to accomplish this:

```
CreateService(
  drive.Contract.Identifier,
     Microsoft.Robotics.Simulation.Partners.CreateEntityPartner(
  "http://localhost/" + robotBaseEntity.State.Name)
);
```

At this point, the child entities, such as the laser rangefinder and bumper sensors, can be created. To do this, you will go through a similar process where you create a new *LaserRangeFinderEntity* object and use the *CreateService* method to specify the laser rangefinder service that will be bound to it. To see a step-by-step example of this, refer to the source code for Simulation Tutorial 5, which can be found in the \samples\SimulationTutorials\Tutorial5 folder.

Creating New Entity Types

MSRS provides implementation classes and entity types to support the Pioneer 3DX, iRobot Create, KUKA LBR3, and LEGO NXT robots. To simulate a new robot or sensor, you will first need to create a new entity type. This can be done by modifying the Entities class file, which is included with the MSRS installation and located in the \samples\simulation\entities folder.

Note If you want to modify the code for an existing entity type, you will need to copy and paste the code for that entity type and create a new entity type.

Summary

- The simulation service included with MSRS allows you to create computer simulations that involve simple or complex objects. These simulations can be very useful when multiple people need to experiment with expensive robots or there is a need to prototype a new robotics hardware design.

- The VSE allows programmers to build simulations easily by adding new entities using one of the entity types provided with MSRS. Entities can be simple, such as spheres, boxes, and capsules, or they can be complex, such as robotics hardware.
- To use VSE, you will need a graphics card that is compatible with DirectX 9 and supports the following pixel and vertex shader versions: PS_2_0 and VS_2_0 or higher.
- MSRS includes several simulations that demonstrate the basics involved with creating and running simulations. The source code for each of these simulations is available in the \samples\SimulationTutorials folder and can be used as a starting point for creating custom simulations.
- Simulations can be created programmatically by creating a new DSS service and adding the simulation engine as a partner service. New entities are added using the entity types provided in the Entities class file. New entity types can be created by modifying this class file.

Chapter 5
Remote Control and Navigation

Up until this point, much has been covered about the basics of Microsoft Robotics Studio (MSRS) and the tools it supplies. It was necessary to do this for you to understand all that is possible with MSRS. This chapter will dive down deep into the code and look at what you need to do to operate your robot. In this chapter, we will focus on creating a Decentralized Software Services (DSS) service for the iRobot Create robot. We will use drive services included with MSRS to operate the robot. The service that we create will receive notifications whenever the motor is enabled or the bumper is pressed.

Working with the iRobot Create

The iRobot Create (see Figure 5-1) is a fully assembled robot that is designed for use by educators and robotics hobbyists. It is based on the popular vacuuming robot: the Roomba by iRobot. For more information about the Roomba and iRobot, see the following URL: *http://store.irobot.com*. To view a list of the available Roomba models, click on the Robots link, and then click Vacuum Cleaning.

Figure 5-1 The iRobot Create is a fully assembled robot designed for programmers interested in working with robots.

The Create does not vacuum your floors, but it does offer more options for expansion. For example, MSRS developers added cameras to multiple Create robots and used them to compete in a sumo robot competition at the 2007 Microsoft Mobile and Embedded DevCon (MEDC) conference. For more information on the Create robot and how to purchase one, go to the iRobot store Web site (*http://store.irobot.com*), click the Robots link, and then click Programmable.

Note You can get information on how to assemble your own sumo bot by going to the MSDN Web site at *http://msdn.microsoft.com* and searching the phrase "sumo robot assembly instructions." Additionally, the MSRS Web site includes a simulation that represents the sumo bot competition. You can download this simulation by going to the Microsoft Download Center Web site at *http://www.microsoft.com/downloads* and searching the phrase "sumo competition."

The nice thing about using the Create is that no assembly is required, unless you want to add sensors beyond the 32 already available. You will need to do some configuration (refer to the section titled "Configuring the iRobot Create" in Appendix B), but it is relatively minor compared with some of the other supported robots. This makes it a good hardware platform for students who are interested in learning about programming robotics applications but not about the actual hardware itself.

Tip In September 2007, MSRS released a set of tutorials that use the Create robot with Visual Programming Language (VPL) applications to demonstrate programming techniques. You can download these tutorials by going to the Microsoft Download Center at *http://www.microsoft.com/downloads* and searching the words "introductory courseware robotics."

It is just this type of instant accessibility that makes pre-assembled robots like the Create of interest to organizations like the Institute for Personal Robots in Education (IPRE). (See the sidebar titled "Profile: Institute for Personal Robots in Education.") Although the IPRE does not use the Create, it does use a smaller, pre-assembled robot named the Scribbler to teach computer science concepts.

Profile: Institute for Personal Robots in Education

The Institute for Personal Robots (IPRE) was founded by Georgia Tech and Bryn Mawr College (see *http://www.ipre.org*). The institute seeks to take away the complexity of robotics and introduce the field to a wider audience. One way it does this is through the use of simple hardware and software platforms. The institute uses a pre-assembled robot made by Parallax named Scribbler. Along with a simple programming language named MyRobotics (MyRo), the Scribbler can be used to teach students basic computer science concepts in a fun and approachable way.

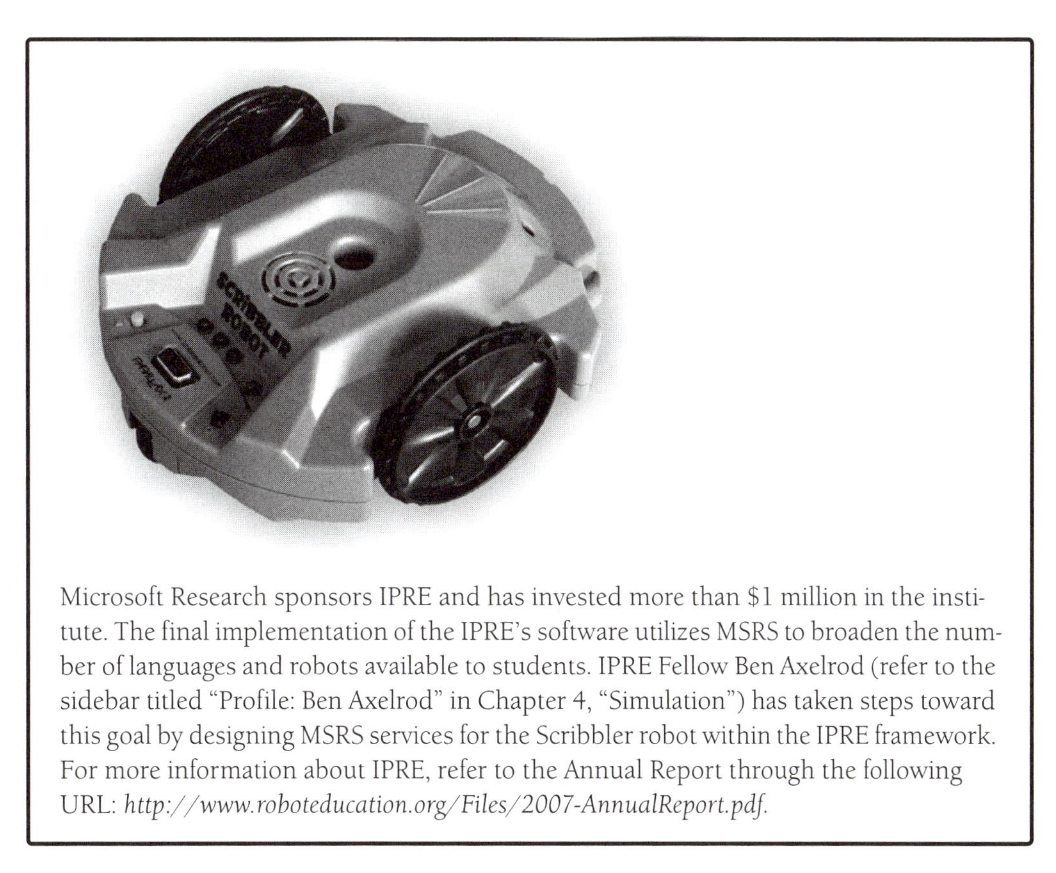

Microsoft Research sponsors IPRE and has invested more than $1 million in the institute. The final implementation of the IPRE's software utilizes MSRS to broaden the number of languages and robots available to students. IPRE Fellow Ben Axelrod (refer to the sidebar titled "Profile: Ben Axelrod" in Chapter 4, "Simulation") has taken steps toward this goal by designing MSRS services for the Scribbler robot within the IPRE framework. For more information about IPRE, refer to the Annual Report through the following URL: *http://www.roboteducation.org/Files/2007-AnnualReport.pdf.*

Building a BasicDrive Program

In this section, we will walk through the steps for creating a simple service that is used to drive your Create robot. You can use this same service to control any robot with a two-wheel differential drive system and at least one contact sensor. To use this same service when operating a different robot, you need to change only the manifest file you load with the service.

The service created in this chapter is based on code included in Robotics Tutorials 1 and 4. Additionally, the service will alert the user, using a Windows dialog box, that a bumper has been pressed. The text in this chapter will cover, in great detail, all the steps required to build this type of generic service. If you do not wish to create a Visual Studio project yourself, you can look at the C# code provided on the book's companion Web site.

Note If you have not yet read Chapter 2, "Understanding Services," you should do so now. Chapter 2 explains many of the fundamentals regarding a service that will now be put into practice.

The BasicDrive service assumes that you want to drive a single robot. It includes five buttons: four represent each direction the robot can move, and one makes the robot stop. The service also subscribes to the bumper sensor, and, whenever you press the bumper, the Utility service is called upon to send a message to a Windows alert dialog box.

To begin, you need to create a new DSS service using the Visual Studio template installed with MSRS. To do this, open Visual Studio 2005 and create a new project named BasicDrive using the Simple Dss Service (1.5) template (see Figure 5-2).

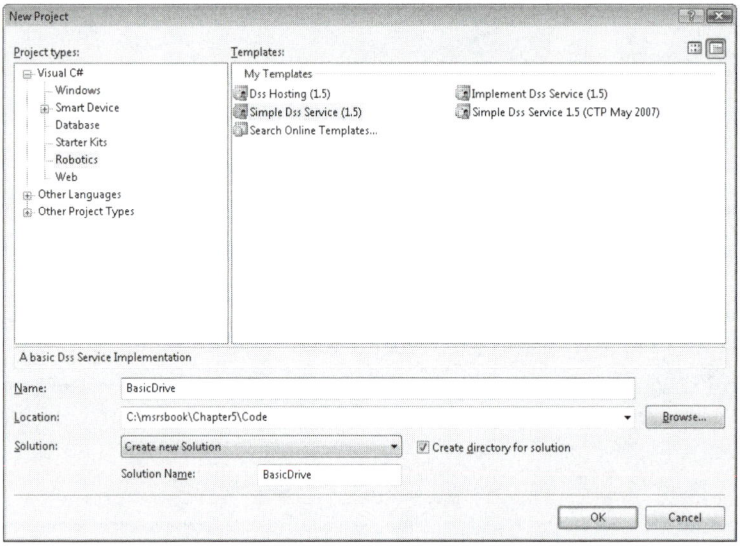

Figure 5-2 Create a new DSS service using the Visual Studio project template installed with MSRS.

> **Tip** When you create a new project, Visual Studio asks you to specify the project location. By default, Visual Studio points to a location within \Documents\Visual Studio 2005\Projects. To run the BasicDrive service, you execute the DssNode program as an external program and load two manifest files. One of these manifest files is for the Create robot, and this file resides within the MSRS application directory. For the application to locate this file properly, you need to save your BasicDrive project to a location within the MSRS application directory, such as the following: ...\samples\BasicDrive. (If you installed MSRS to the default application directory, you will find this directory in the path C:\Microsoft Robotics Studio (1.5)\samples.)

Add Windows Form

The interface for this project is a Windows form that includes controls you use to drive your robot. By default, a Windows form is not included with the project created using the Simple Dss Service template. To begin, you need to add a new Windows form to the BasicDrive service project by right-clicking the project in Solution Explorer and selecting Add and then Windows Form. Name the new form DriveDialog and use either a .cs or .vb file extension. Click Add.

Change the size of the form by right-clicking on the design surface of the form and selecting Properties. Make the following change to the form properties (the property values are the same regardless of whether you are creating a Visual Basic or C# project): Size= 202,161.

You can then add five buttons to the design surface by dragging the controls from the Toolbox. After you add the buttons to the form, you can change their properties by right-clicking each control and selecting Properties. Make the following changes to the button controls (the following property values also position the controls properly):

- **Button 1** Name=btnRight

  Font Name=Marlett, Font Size=9.75pt, Font Unit=Point, Font Bold=True, Font GdiCharSet=2, Font GdiVerticalFont=False, Font Italic=False, Font Strikeout=False, Font Underline=False

  Location=117,50

  Size=35,23

  Text=4

- **Button 2** Name=btnLeft

  Font properties=same values used for Button 1

  Location=36,50

  Size=35,23

  Text=3

- **Button 3** Name=btnForward

  Font properties=same values used for Button 1

  Location=77,20

  Size=35,23

  Text=5

- **Button 4** Name=btnBackwards

  Font properties=same values used for Button 1

  Location=77,80

  Size=35,23

  Text=6

- **Button 5** Name=btnStop

  Font properties=same values used for Button 1

  Location=77,50

  Size=35,23

  Text=1

Modify the Constructor

After you add all the controls and set the values for the properties, the form should resemble Figure 5-3. The next step is to modify the forms constructor so that it can accept messages sent from the service. You do this by right-clicking on the form design surface and clicking View Code. Replace the class definition for the form with the following code:

```
public partial class DriveDialog : Form
{
      BasicDriveOperations _mainPort;

      public DriveDialog(BasicDriveOperations mainPort)
      {
          _mainPort = mainPort;

          InitializeComponent();
      }
}
\
```

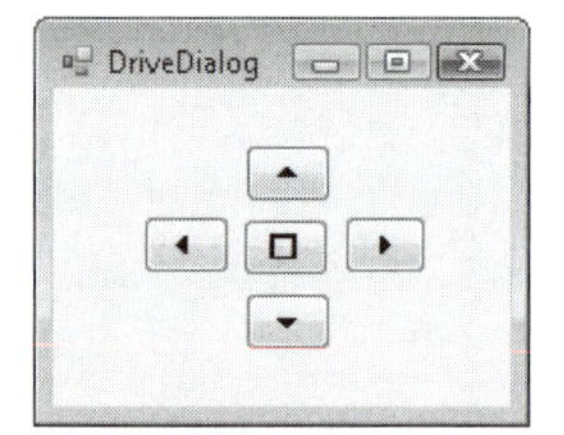

Figure 5-3 The DriveDialog represents a simple service that you can use to drive a single robot using the button controls. It also initiates a Windows dialog box whenever a bumper is pressed.

In the previous code snippet, we declared an instance of the *BasicDriveOperations* class, and we created a new variable named *_mainPort*. We then added this variable as a parameter to the constructor for the *DriveDialog* form. Doing this allows messages to be passed from the Windows form to the service through the main port.

Before we can add the code that executes when a user clicks one of the buttons, we need to add supporting code to the service project. We will do this in the section titled "Add Code to the Implementation Class," later in this chapter.

Defining the Service Contract

When you create a DSS service using the built-in template, contract and implementation class files are automatically created for you. For this service, the contract class file is named BasicDriveTypes.cs. This is where you place type declarations for the state variables used by this service. It is also where you define what DSS operations are allowed for this service.

If you are following along with the book and have created your own BasicDrive project, then you will notice a class named *Contract*, which contains a variable named *Identifier*. This variable was added to the project automatically when you created it using the template. The value for this variable represents the service contract for this service. The service contract contains the information that other DSS services need to communicate with your service, and, by default, it is assigned a unique name that includes the month and year that you created the service. For example, the following identifier was assigned to the BasicDrive service because it was created in September 2007:

```
public const String Identifier = "http://schemas.tempuri.org/2007/09/basicdrive.html";
```

Add State Variables

The state represents the service at the time it is requested, and state variables are containers for the individual pieces of information that make up the state. For this service, we will have a single state variable that indicates whether the motor has been enabled. To add this state variable, append the following code to the public class named *BasicDriveState* (this class definition was created automatically by the Visual Studio template and is included in the BasicDriveTypes class file):

```
/// <summary>
/// Indicates if the Motor is running.
/// </summary>
[DataMember]
public bool MotorRunning
{
  get { return _motorRunning; }
  set { _motorRunning = value; }
}
```

The value of the state variable changes as the user presses buttons on the Windows Form. Notice that the public declaration is preceded by a *DataMember* attribute. All public fields or properties that represent the service state must include the *DataMember* attribute. This ensures that the variable is included with the proxy version of the service assembly. Without this attribute, the public field is not available to the service. This means that, if you use a Web browser to get the state for a service, the Web browser returns only public state variables marked with the *DataMember* attribute.

Add New Service Operations

Each service is defined by the data that it consumes or exposes. A service is also defined by the operations it can perform. Service operations are responsible for performing specific actions, such as getting or replacing the service state. They can also be used to perform actions specific to a service. For example, the BasicDrive service needs an operation capable of issuing a request to move the robot in a certain direction. The service operations required for the BasicDrive service include the following:

- **DsspDefaultLookup** Defines a handler for a Lookup message, which is used to return the service context
- **DsspDefaultDrop** Allows the service to support the drop message
- **Get** Used to return the current state for the service
- **Move** Used to move the robot in a certain direction
- **Replace** Replaces the value of all state variables associated with the service

Each new service operation must be associated with a class definition. This is what determines how the code associated with the operation is implemented. It defines the data for your class. MSRS includes default interfaces for the *DsspDefaultLookup* and *DsspDefaultDrop* operations, so you do not need to add code to support them. Additionally, code to support the *Get* operation is included with the Visual Studio template.

You need to add code that supports the *Move* and *Replace* operations. The code to represent the *Move* and *Replace* operations is shown as follows (you should add this code block beneath the public class named *Get*):

```
[DataContract]
public enum MoveType
{
    Stop,
    Left,
    Right,
    Forward,
    Backward
}

[DataContract]
public class MoveRequest
{
    [DataMember]
    public MoveType Direction;
    public MoveRequest()
    {
    }

    public MoveRequest(MoveType direction)
    {
        this.Direction = direction;
    }
}


public class Move : Submit<MoveRequest,
    PortSet<DefaultSubmitResponseType, Fault>>
```

```
{
    public Move()
    {
    }

    public Move(MoveRequest body)
        : base(body)
    {
    }
}
public class Replace : Replace<BasicDriveState,
     PortSet<DefaultReplaceResponseType, Fault>>
{
     public Replace()
     {
     }

     public Replace(BasicDriveState body) : base(body)
     {
     }
}
```

In the preceding code block, the class named *Move* derives from the generic type *Submit*. The *Submit* method, which is part of the DSS Service model, handles messages and accepts two parameters: one for the message body and one for the message response. In this case, the output for the public method named *MoveRequest* represents the body of the message. If the *Move* operation is successful, the response can be passed back using the default *Submit* response type. Otherwise, a fault message can be passed back.

Note The interfaces defined in the *DriveDialogTypes* class file are just structure definitions. They do not include the code that executes when an operation is performed. You will add that code in the section titled "Add Code to the Implementation Class," later in this chapter.

The *MoveRequest* is defined in the class named *MoveRequest*. The class definition is preceded by the *DataContract* attribute, which is necessary for the request to be used by an operation. The request includes a variable named *Direction*. This variable is preceded by the *DataMember* attribute, which is necessary for the variable to be exposed to the service. The *Direction* variable is used to store the direction in which the robot should move. The value for this variable can be one of the values specified in the *MoveType* enumeration.

Tip If you are building your own service by stepping through the code, then you should stop and do a project build before continuing on to the next section. You can build the project by clicking File and then Save All. Next, click Build and then Build Solution. Ensure that the status bar in the bottom left corner indicates that the build is successful. If it is not, go back now and determine where the error occurred.

Modify the PortSet

The PortSet includes a list of DSS operations that can be performed by the service. By default, the Visual Studio template includes three operations: *DsspDefaultLookup*, *DsspDefaultDrop*, and *Get*. This is found in the public class declaration named *BasicDriveOperations*. To build the BasicDrive service, you need to locate the code for the BasicDrive main operations port and modify the class declaration to include two new operations (*Move* and *Replace*) as follows:

```
[ServicePort]
public class BasicDriveOperations : PortSet<DsspDefaultLookup,
                                   DsspDefaultDrop,
                                   Get,
                                   Move,
                                   Replace>
{
```

Add Code to the Implementation Class

In this project, the implementation class is named BasicDrive.cs. The implementation class is where the majority of your service code will reside. It contains the code that reads incoming data from sensors and sends commands to operate the robot's motors.

Set References

The first thing to do is include references to assemblies needed by this project. To do this, right-click the References folder in Solution Explorer and click Add Reference. From the Add Reference dialog box (see Figure 5-4), scroll through the components listed in the .NET tab and locate the following (select each assembly while holding down the Ctrl key and click OK to add):

- **Ccr.Adapters.Winforms** This assembly allows your Windows forms application to run under Concurrency and Coordination Runtime (CCR). This is necessary because the form will need to be launched from our BasicDrive service.
- **RoboticsCommon.Proxy** This assembly, which is included with MSRS, allows you to access a variety of namespaces used to control a robot. In this chapter, we add a reference to the *Microsoft.Robotics.Services.Drive.Proxy* namespace. This gives you access to drive operations used to operate a two-wheel differential drive system. It also includes the *Microsoft.Robotics.Services.ContactSensor.Proxy* namespace, which allows you to receive feedback from a contact sensor.
- **Utility.Y2006.M08.Proxy** This assembly provides access to several supporting functions, such as launching a Windows dialog box, performing math functions, generating sounds from your desktop, and launching a URL. In this chapter, we utilize the *Microsoft.Robotics.Services.Sample.Dialog.Proxy* to launch an Alert dialog box whenever the bumper is pressed.

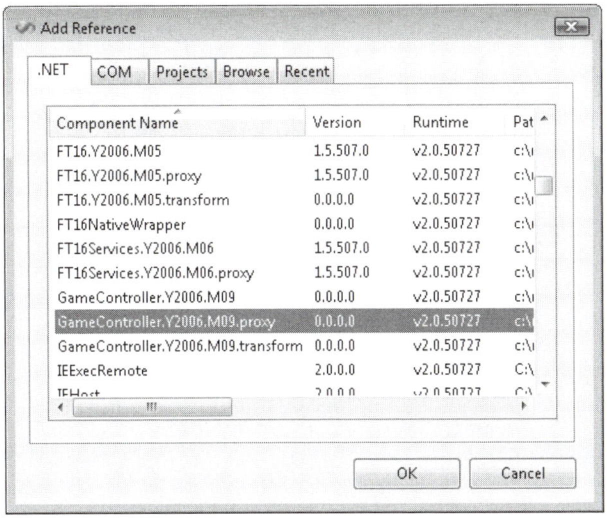

Figure 5-4 The Add Reference dialog box is used to add references to external assemblies.

> **Tip** After you add the references to your project, you can view a list of available namespaces and functions using the Object Browser. This is one way of learning what you can do with the assemblies included with MSRS. To get to the Object Browser, click View, and then click Object Browser from the menu. Alternatively, you can use the DssInfo command-line tool that was presented in Chapter 2.

For each reference you added, you need to modify the *Copy Local* and *Specific Version* properties for that reference. You should be able to access the properties for each reference in the bottom right pane. (If necessary, right-click on a reference and choose Properties to display the Properties dialog box for that reference.) Set both properties to a value of False. This ensures that copies of the assemblies are not copied to the local code directory. This helps to prevent any errors that might occur if the reference is later changed.

After you add the references, you need to add namespace references to the top of the BasicDrive.cs class file. The statements, as follows, should appear before the namespace declaration:

```
using Microsoft.Ccr.Adapters.WinForms;
using drive = Microsoft.Robotics.Services.Drive.Proxy;
using bumper = Microsoft.Robotics.Services.ContactSensor.Proxy;
using dialog = Microsoft.Robotics.Services.Sample.Dialog.Proxy;
```

> **Note** It is not necessary to include an alias with your namespace declarations. The benefit to using the prefix is that you do not have to reference the full namespace in your code, thus making the code easier to read.

In addition to the namespace references listed previously, you need to add a namespace reference for the *W3C.Soap* namespace. This is a member of the *DssBase* assembly, and it is used by the BasicDrive service to access the *Fault* object. Add the following code below the other namespace references:

```
using W3C.Soap;
```

Tip When you created the service using the Visual Studio template, it added attributes for the display name and service description in the BasicDrive class file. The attributes appear in the code as follows:

```
[DisplayName("BasicDrive")]
[Description("The BasicDrive Service")]
```

The text within these attributes is displayed in Control Panel when the service is published to the local service directory. By default, the Display Name is the name of the project, and the Description also includes this name. It is a good idea to add more useful text to the description. There is no significant limit to the number of characters this attribute can contain.

You can also add the description to all classes marked with the *DataContract* attribute and properties marked with the *DataMember* attribute. The description appears in the information listed by the DssInfo command-line utility and also in the tool tips displayed in VPL.

Add Partnerships

A partner is the mechanism that allows one service to use the data from another service. The services are bound together through their service contracts. Partnerships provide a way to link services together. This is a great way to reuse code, and MSRS provides several assemblies containing functions you can use in your service. The *Robotics.Common.Proxy* assembly contains several namespaces that you can use to operate your robot. This can include everything from managing sensors to driving the robot. By referencing the *Microsoft.Robotics.Services.Drive.Proxy* namespace using the drive prefix, you will be able to easily reference the drive functions included within this namespace. To add the partnership, add the following code directly below the declaration for the main port:

```
[Partner("Drive", Contract = drive.Contract.Identifier, CreationPolicy =
PartnerCreationPolicy.UseExisting)]
private drive.DriveOperations _drivePort = new drive.DriveOperations();
private drive.DriveOperations _driveNotify = new drive.DriveOperations();
```

In the previous code snippet, private variables were created to reference two ports, named _drivePort and _driveNotify. Commands to move or turn the robot can be sent to the port named _drivePort. The _driveNotify port is used to receive notifications from the drive service. The notifications are similar to events, and, for the drive service, they are used to determine when the motor is enabled.

You need to establish a partnership with the Generic Contact Sensors service. This partnership declaration is similar to the one created for the drive service. You also need to define two ports, where one port is the bumper port, and the other receives notifications whenever a bumper is pressed. The code for this declaration is shown as follows (you should include it below the other Partner declaration):

```
[Partner("bumper", Contract = bumper.Contract.Identifier,
     CreationPolicy = PartnerCreationPolicy.UseExisting)]
private bumper.ContactSensorArrayOperations _bumperPort = new
     bumper.ContactSensorArrayOperations();
private bumper.ContactSensorArrayOperations _bumperNotificationPort = new
     bumper.ContactSensorArrayOperations();
```

Tip Each supported robot provides a service to access the contact sensors for that robot. If you are building a service intended to work only with a specific robot, it is best to use the contact sensor service for that robot because it includes additional functionality not available with the generic contact service.

The final partnership to establish is one with the Simple Dialog service, which is a member of the *Utility.Y2006.M08.Proxy* assembly. This partnership allows you to send a message to a Windows alert dialog box. For the BasicDrive service, an alert dialog box is initiated every time the robot's bumper is pressed. The code for this declaration is shown as follows:

```
[Partner("SimpleDialog", Contract = dialog.Contract.Identifier, CreationPolicy =
PartnerCreationPolicy.UsePartnerListEntry)]
dialog.DialogOperations _simpleDialogPort = new dialog.DialogOperations();
```

Modify the *Start* Method

When you create a service using the MSRS template for Visual Studio, a method named *Start* is created automatically. This method is called when the service starts, and it is where you add initialization code specific to your service. For the BasicDrive service, you need to replace the code in the existing *Start* method with the following:

```
protected override void Start()
{
   //Initialize the state
   state.MotorRunning = false;

   // Used to publish the service to the local service directory
// The Start method will also activate DSSP operation handlers that
// are used to listen on the main port for requests and call
// the appropriate handlers
base.Start();

// Invoke the Windows Dialog using a delegate which acts as a pointer
   //  to the Windows form.
   WinFormsServicePort.Post(
```

```
        new RunForm(
           delegate()
           {
               return new DriveDialog(
                ServiceForwarder<BasicDriveOperations>(ServiceInfo.Service)
               );
           }
        )
     );

    //Register the message handlers with the main interleave
    Activate(Arbiter.Interleave(
      new TeardownReceiverGroup
      (
         Arbiter.Receive<DsspDefaultDrop>(false, _mainPort, DropHandler)
      ),
      new ExclusiveReceiverGroup
      (
         Arbiter.ReceiveWithIterator<Replace>(true, _mainPort,
            ReplaceHandler),
         Arbiter.ReceiveWithIterator<Move>(true, _mainPort, MoveHandler)
       ),
       new ConcurrentReceiverGroup
       (
         Arbiter.Receive<drive.Update>(true, _driveNotify,
           NotifyDriveUpdate),
         Arbiter.ReceiveWithIterator<Get>(true, _mainPort, GetHandler),
         Arbiter.Receive<bumper.Update>(true, _bumperNotificationPort,
           SensorHandler)
       )
    ));

// Request that a notification is sent to the _driveNotify port whenever
// an update occurs since this means the motor is enabled or disabled
_drivePort.Subscribe(_driveNotify);
  // Subscribe to the generic contact sensors service so we
  // will be notified each time a bumper sensor is pressed
  _bumperPort.Subscribe(_bumperNotificationPort);

}
```

The first thing the *Start* method does is initialize any state variables by assigning them a value. For this service, the only state variable is named *MotorRunning*, and we always initially set the value to False.

The *Start* method should already include a call to another *Start* method, which is part of the Decentralized Software Services Protocol (DSSP) service base. This method is called when the service initialization is complete, and it is responsible for activating DSSP operation handlers. The DSSP operation handler listens on the main port and calls the appropriate message handlers whenever a DSS operation is posted to that port. It also publishes the service to the local service directory so that it appears in the Control Panel page for MSRS. You only need to keep the code "**base.Start();**" to allow this to happen.

The next statement is a call to the *Post* method, which is part of the WinForms service. This is where you invoke the Windows form associated with this service. The Windows form, which we created in an earlier step, is the interface for operating the robot. In this code, the *Service-Forwarder* method is used to pass a strongly typed PortSet to the Windows form. This is what allows the Windows form to communicate with the BasicDrive service.

Message handlers are registered with the main interleave. The interleave assigns incoming tasks from these handlers to an internal queue. It then manages the execution of these tasks according to what group they were assigned. If the handlers are assigned to the concurrent group, they run at the same time. If the handlers are assigned to the exclusive group, the tasks within those handlers wait for all other tasks to complete first. For the BasicDrive service, the handlers associated with the *Replace*, *Move*, and *Drive Update* operations are assigned to the exclusive group. This is because these handlers involve changing the service state. The remaining handlers, associated with the *Get* and *Bumper Update* operations, are assigned to the concurrent group.

You need to create two subscriptions for this service. The first subscription is for the drive service, and the code in the *Start* method specifies that a notification should be sent to the drive notify port every time an update occurs. In this case, the *Arbiter* class, which is part of the CCR, is used to create the subscription. By using the *Arbiter* class, a thread does not have to wait until a notification is received before continuing to the next line of code. Specifically, the Receive routine is used to initiate a single item receiver. This means that the receiver waits for a single message before calling the handler. The first parameter indicates whether the receiver continues listening for requests.

The other subscription is for the generic contact sensors service. This subscription also uses the Receive routine to initiate a single item receiver. In this case, the update is for the contact sensor, and this is used to tell us when the bumper is pressed or released.

Create Message Handlers

Each service operation listed in the contract class has a message handler included in the BasicDrive.cs class file. The code within the message handler is executed whenever the associated operation is called. The *Move* operation, which was defined in the section "Add New Service Operations" earlier in this chapter, is called upon whenever the user wants to move the robot in any direction. It uses the *Direction* property, which is part of the incoming message body, to determine in which direction it should move. For example, the following code, which you should add to the implementation class, is what the message handler named *Move-Handler* should look like:

```
public virtual IEnumerator<ITask> MoveHandler(Move move)
{
     // Check to see if the motor is already running. It is not,
     // then we need to call the EnableMotor function
     if (!_state.MotorRunning)
     {
```

```
        yield return EnableMotor();
        _state.MotorRunning = true;
    }

    // Initialize the drive request that will be used to send
    // a request to the robot through the main port
    drive.SetDrivePowerRequest request = new drive.SetDrivePowerRequest();

    string errorMessage = "";
    //Determine which direction the robot should move toward
    switch (move.Body.Direction)
    {
         case MoveType.Backward:
            //Sending the same negative values causes it to
            //move backwards
            request.LeftWheelPower = -0.5;
            request.RightWheelPower = -0.5;
            errorMessage = "Failed to move the robot backwards";
            break;
         case MoveType.Forward:
            //Sending the same positive value causes it to
            //move forwards
            request.LeftWheelPower = 0.5;
            request.RightWheelPower = 0.5;
            errorMessage = "Failed to move the robot forwards";
            break;

         case MoveType.Left:
            // differing power values which cause the robot to turn in
            // one direction or the other
            request.LeftWheelPower = 0;
            request.RightWheelPower = 0.5;
            errorMessage = "Failed to turn the robot to the left";
            break;
         case MoveType.Right:
            // alternating power values which cause the robot to turn in
            // one direction or the other
            request.LeftWheelPower = 0.5;
            request.RightWheelPower = 0;
            errorMessage = "Failed to turn the robot to the right";
            break;
         case MoveType.Stop:
            // sending no power causes the robot to stop
            request.LeftWheelPower = 0;
            request.RightWheelPower = 0;
            errorMessage = "Failed to stop the robot";
            break;
     }
```

The *MoveHandler* method checks to see if the state variable named *MotorRunning* is set with a value of True. If it is set with any other value, this means the motor is disabled, and it needs to be enabled before continuing. To allow for this, you need to add a function named *EnableMotor*. This function, which is shown below, uses other functions from the *Microsoft.Robotics.Services.Drive.Proxy* namespace to enable the motor:

```
private Choice EnableMotor()
{

    // This is where we send the request to the port, but we use the
    // Arbiter.Choice function to do this so that we either enable the
    // drive and return a response, or we return a failure and log the
    // resulting error message
    return Arbiter.Choice(
        _drivePort.EnableDrive(true),
        delegate(DefaultUpdateResponseType response) { },
        delegate(Fault fault)
        {
            LogError(null, "Failed to enable motor", fault);
        }
    );
}
```

The *EnableMotor* method sends an update message to the drive service requesting that the motor is enabled. The body for this message is the method call to _drivePort.EnableDrive(). The *EnableDrive* function sends the actual message, and it is called inside an *Arbiter.Choice* function. The *Arbiter* class is part of the CCR. It provides several helper functions that allow you to coordinate asynchronous services. One of these functions is named *Choice*, and it guarantees that only one choice executes. Even if the service fails to respond and returns both a success and failure, only the first return message will be processed. In the *EnableMotor* function, *Arbiter.Choice* is used to send the enable motor request to the _drivePort. This ensures that either the motor is enabled or an error is logged.

After the motors have been enabled, the next step in *MoveHandler* is to determine in which direction the robot needs to move. You do this by using a switch statement to inspect the *Direction* property. The *Direction* property is passed to the *MoveHandler* through the message body. It should contain one of the values defined in the *MoveType* enumeration (Forward, Backward, Right, Left, or Stop).

A drive power request must be created to move or stop the robot. A two-wheel differential drive system works by sending alternating power levels to the motors driving each wheel; in this case, there are two motors and two wheels. You can use the *SetDrivePowerRequest* class to create the request for adjusting these power levels. The class uses two properties, *LeftWheelPower* and *RightWheelPower*, which indicate power settings for each wheel.

Each power level property can accept a double value anywhere from −1.0 to 1.0. Positive values are used for driving forward and negative values for driving backward. To drive forward, you need to assign any value between 0 and 1 to both the right and left wheels. If you want the robot to turn left or right, then the value assigned to the left and right wheels must differ. For example, to turn left, you could assign a value of 0 to the left wheel and a value of 0.5 to the right wheel. If the values assigned to each motor are reversed, then the robot spins on the center of its wheelbase. For example, you would not want to use values of -0.5 and 0.5 for the left and right wheels.

The final task for the *MoveHandler* is to send the request to the drive port. Just like we did for the *EnableMotors* function, we use the *Arbiter.Choice* function. The request is sent to the drive port, and, if a failure occurs, the error is logged.

In addition to the handlers associated with each DSS operation specified in the PortSet, you need to add message handlers for the subscriptions that we created in the *Start* method. The BasicDrive service initiates two subscriptions: one to receive notifications when the motor is enabled and the other receives notifications when a bumper is pressed. The name of the handler method for the drive service is *NotifyDriveUpdate*. This method is responsible for updating the value of the *MotorEnabled* state variable. It does this by sending a new Replace request to the main port. You will need to add code for this method, which is seen as follows:

```
private void NotifyDriveUpdate(drive.Update update)
{
     BasicDriveState state = new BasicDriveState();
     state.MotorEnabled = update.Body.IsEnabled;

      _mainPort.Post(new Replace(state));
}
```

You also need to add code below the other handler methods, for the handler method named *SensorHandler*. This method is called every time the bumper state is updated. The *SensorHandler*, which is shown as follows, creates an alert request and sends it to the simple dialog port whenever the notification indicates that a button was pressed:

```
private void SensorHandler(bumper.Update notification)
{
    if (notification.Body.Pressed)
    {
         dialog.AlertRequest request = new dialog.AlertRequest();
         request.Message = @"The bumper was pressed";
         _simpleDialogPort.Alert(request);

    }
}
```

Add Code to Click Events

After you have added all the supporting code, you can return to the Windows form and add code for the click events. A click event occurs each time a user clicks one of the buttons on the Windows form, and this is where we link the form to the service. The code in the click event posts an applicable DSS operation to the main port.

To add the code, right-click the DriveDialog form in Solution Explorer, and then click View Code. Add the following code below the form constructor:

```
private void btnRight_Click(object sender, EventArgs e)
{
     mainPort.Post(new Move(new MoveRequest(MoveType.Right)));
```

```
}

private void btnForwards_Click(object sender, EventArgs e)
{
    _mainPort.Post(new Move(new MoveRequest(MoveType.Forward)));
}

private void btnLeft_Click(object sender, EventArgs e)
{
    _mainPort.Post(new Move(new MoveRequest(MoveType.Left)));
}

private void btnBackwards_Click(object sender, EventArgs e)
{
    _mainPort.Post(new Move(new MoveRequest(MoveType.Backward)));
}

private void btnStop_Click(object sender, EventArgs e)
{
    _mainPort.Post(new Move(new MoveRequest(MoveType.Stop)));
}
```

The last thing to add is a clean-up method for handling when the form is closed. In this case, we want to send a request that drops the BasicDrive service. The code for this method, which you can add below the other click events, is as follows:

```
protected override void OnClosed(EventArgs e)
{
    _mainPort.Post(new DsspDefaultDrop(DropRequestType.Instance));
    base.OnClosed(e);
}
```

Note If you are building your own service by stepping through the code, then you should stop and do a project build before continuing on to the next section. You can build the project by first clicking File and then Save All. Next, click Build and then Build Solution. Ensure that the status bar in the bottom left corner indicates that the build was successful. If it was not, go back now and determine where the error(s) occurred.

Change the BasicDrive Manifest

If you tried to run the BasicDrive service now, you would receive a message in the command window stating that the partner enumeration failed and the service would not start. This is because the BasicDrive service uses partners that must be coordinated to work properly. When you declare a new partner, you use an attribute named *CreationPolicy* to specify how the service should be started. You can set the *CreationPolicy* attribute with one of the following values:

- **CreateAlways** Always creates a new instance of the service.
- **UseExisting** Uses an existing instance of the service from the local directory.

- **UseExistingOrCreate** First checks to see if the service is running on the local directory. If it is not, it attempts to create it.
- **UsePartnerListEntry** References the *PartnerListEntry* attribute included in the manifest file. This can be used to specify contracts for one or more associated services.

When you have a service that partners with other services, such as the BasicDrive service, there needs to be some method of coordinating how the runtime starts these services to avoid conflicts. The easiest way to do this is through the use of a manifest file. You use the manifest file to specify a list of services associated with the base service. Even though this file is XML-based and, thus, can be modified with a text editor, it is best that you use the DSS Manifest Editor tool instead.

The DSS Manifest Editor tool (see Figure 5-5) allows you to build a manifest by dragging and dropping services onto a design surface. To resolve the potential problem with coordinating the BasicDrive service, you need to open the Microsoft DSS Manifest Editor, which is located in the MSRS menu folder. This opens a new manifest file, and, from here, you can locate the BasicDrive service in the left-hand window. Drag an instance of the BasicDrive service onto the design surface, and it should list the Drive, bumper, and SimpleDialog services beneath it. There should be a red error icon next to the SimpleDialog service, which indicates that you have to specify a service for this partnership.

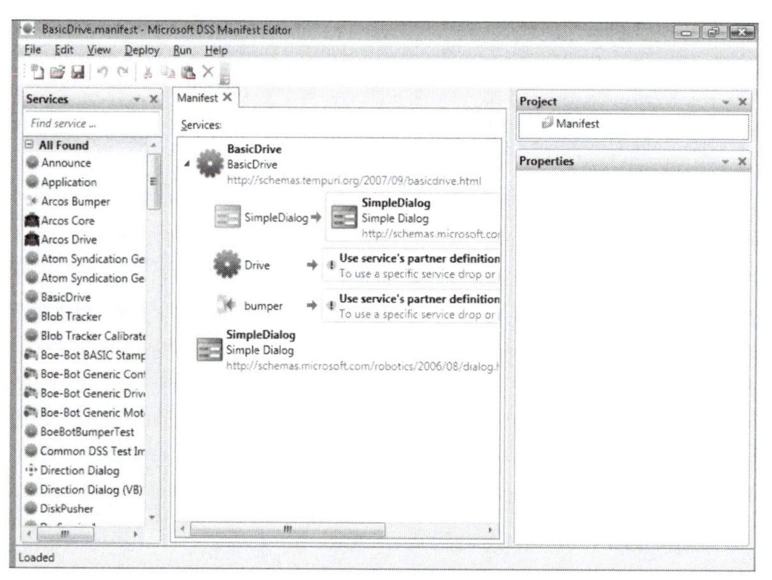

Figure 5-5 You can use the DSS Manifest Editor tool used to easily build service manifest files.

To resolve the red error listed in the Manifest Editor, you need to scroll through the list of services until you find the SimpleDialog service. Drag an instance of this onto the area where the red error icon appears. The final result should appear as it does in Figure 5-5.

Save a copy of the manifest file by clicking File and then Save As. From the Save As dialog box, locate the directory where the BasicDrive service is stored (this should be within the MSRS installation directory). There should be a manifest file named BasicDrive.manifest within this directory. Select this file and click Save. Click Yes if asked to replace the existing file.

Change Project Properties

The BasicDrive service was designed to work with any robot that has a two-wheel differential drive system and at least one contact sensor. However, you still have to indicate which robot will use the service. You do this by loading a manifest file specific to the robot you wish to use. In this chapter, we are working with the Create by iRobot.

The manifest file that is used to access the Create's drive system and contact sensors is named iRobot.DriveBumper.manifest.xml. This file is included with the MSRS installation, and it is located in the \samples\config folder. To load this additional manifest, you need to modify the project properties for the BasicDrive service. You accomplish this by clicking Project and then BasicDrive Properties. Click the Debug tab (see Figure 5-6) and locate the text box named Command Line Arguments. Append the following to the end of this line:

```
-m:"samples\config\iRobot.DriveBumper.manifest.xml"
```

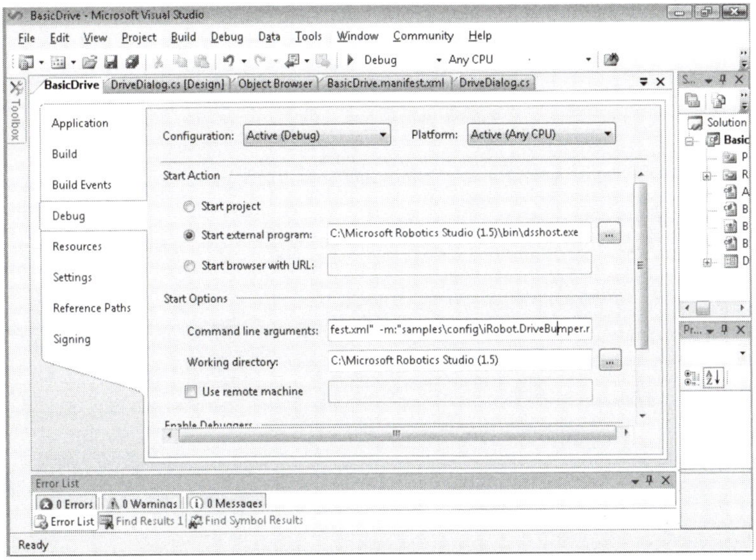

Figure 5-6 You can specify that additional manifest files are loaded when the service is started by appending manifest lines to the command-line arguments within project properties.

After you make the project properties change, you can save your project and rebuild the solution by clicking Build and then Build Solution.

Run the BasicDrive Service

Before running your BasicDrive service, you need to ensure that the Create robot is configured and connected to your development machine. You can connect the Create through a USB cable or a Bluetooth module. If you have not already done so, refer now to the instructions for configuring a Create robot in Appendix B.

Tip If you attempt to execute the service and all you see is a command-line window briefly, but no Windows form, then you need to check the following:

1. You are running the project from a folder within the MSRS installation folder, such as C:\Microsoft Robotics Studio (1.5)\Samples\BasicDrive.
2. The manifest files referenced in the project properties are available in the locations specified.

To run the BasicDrive service from Visual Studio, you need only press Ctrl+F5 to start without debugging or F5 to start with debugging. Either way, you should see a command-line window appear and, within this window, text that indicates the service is being loaded. After the service loads successfully, the Windows form should appear, and you should hear a beep from the robot. At this point, you can move the robot by clicking one of the directional buttons. Use the middle button to stop the robot's movement. If the robot's bumper sensor is pressed, a Windows dialog box should appear with the message "The bumper was pressed."

Summary

- In this chapter, you created a BasicDrive service, which is used to operate the Create by iRobot. The service allows you to drive the robot using directional command buttons. If the robot encounters an obstacle and the bumper is pressed, a Windows dialog box appears with a message stating that the bumper was pressed.
- When you create a new DSS service using the Visual Studio template, a contract and implementation class are created for you. To build the BasicDrive service, you need to add code to these classes and add a Windows form that includes buttons used to drive the robot.
- The BasicDrive service uses one operation named *MoveHandler* to handle all movement requests. This operation will be included in the PortSet, and the message handler named *MoveHandler* will be added to the implementation class.
- When the user clicks one of the direction buttons on the Windows form, a message request using the *Move* operation is sent to the main port. The request includes in which direction the robot should move. This triggers the associated *MoveHandler* message handler to fire and, thus, send commands to the robot instructing it to move in a certain direction.

- Because the BasicDrive service uses multiple partnerships, you must make a change to the manifest file to prevent an error from occurring. The DSS Manifest Editor is the best tool for making changes to the service manifest file.
- Before running the project, make a final change to the project properties for the BasicDrive service. From the Default tab, modify the command-line arguments and append a line loading the manifest specific to the Create robot.

Chapter 6

Autonomous Roaming

In the last chapter, we covered the basic steps for driving a robot using remote control. In this chapter, you have an opportunity to create a service that allows a robot to move around a room unassisted. You will write the service in two versions, and the first version will use a touch sensor to notify the robot of an obstacle collision. In this case, the robot will perform reactively; when a collision occurs, it will back up and move in another direction. The second version of the service will improve upon the first version and utilize a sonar sensor to detect an obstacle before the collision occurs.

In both versions, we will work with one of the Microsoft Robotics Studio (MSRS)–supported robots: the LEGO Mindstorms NXT. This small and affordable robot includes a two-wheel differential drive system and touch, sound, light, and sonar sensors. This easily assembled robot allows you to experiment with the different ways a robot can navigate its environment using the built-in sensors.

Working with the LEGO Mindstorms NXT

LEGO, well known for its line of toys, offers a robotics kit known as LEGO Mindstorms. The latest version of this robot, the NXT, features a 32-bit processor and includes three servo motors and four different sensors. The sensors provided with the NXT include the following:

- **Touch sensor** This simple sensor can function as a bumper for your NXT robot. When the button on the touch sensor is pressed, you can assume the robot has encountered an obstacle and, thus, instruct the robot to either stop or move away from the object. When the button is released, you can assume the robot has cleared the obstacle.
- **Sound sensor** By detecting decibel levels, the sound sensor can determine when the noise level surrounding the robot reaches a certain threshold. This can be useful when there is a need to control the robot using sounds such as loudly clapping your hands or shouting a command.
- **Light sensor** The NXT light sensor is able to detect both reflective and ambient lighting. Reflective lighting occurs when light bounces off an object. The NXT sensor works by using a light-emitting diode (LED) to emit a light. It then uses a receiver to measure the amount of light reflected back. This can be used to detect when the light in a room has gone on or off.

- **Sonar sensor** The sonar sensor for the NXT represents the robots eyes and, interestingly enough, the sensor itself resembles a head with two big eyes. This sensor is used to measure distance in centimeters, and it does this by emitting a high-frequency sound that humans are not able to hear. The sensor works by measuring how long it takes for the sound to echo back. The elapsed time indicates how near the object is to the robot and, thus, can be useful for obstacle avoidance.

The LEGO NXT, which can be purchased online or at local retailers, is a rugged and easily assembled robot that offers an almost endless array of design possibilities. LEGO sponsors a Web site (*http://mindstorms.lego.com/nxt/Overview/default.aspx*) that provides extensive information about the NXT. The Web site also includes access to an online forum and user community that includes robotics hobbyists of all levels. Do not think that the NXT is simply a toy. It offers an extensible and durable robotics platform that you can use to perform many common robotics tasks.

> **Tip** Several third-party vendors offer additional sensors for the LEGO NXT. One such company, Vernier Software and Technology, offers a sensor adapter for the LEGO NXT (refer to *http://www.vernier.com/nxt/*). The adapter can be used to add one of 32 sensors, such as the Accelerometer, Barometer, Magnetic Field Sensor, Colorimeter, pH Sensor, Relative Humidity Sensor, Soil Moisture Sensor, Sound Level Meter, Temperature Probe, UVA Sensor, and UVB Sensor. The Vernier Web site hosts several movies that showcase what people are doing with the LEGO NXT and Vernier sensors. The NXT services available with MSRS are extensible so that these other sensors can be added.

The example code in this chapter includes two versions: The first version uses the touch sensor to move the robot after an obstacle collision occurs. The second version uses the sonar sensor to detect obstacles before a collision and, thereby, allows the robot to avoid potential damage. Before you begin, you will need to assemble and configure the LEGO NXT. Refer to the configuration instructions in the section titled "Configuring the LEGO Mindstorms NXT" in Appendix B, "Configuring Hardware."

The LEGO NXT allows you to assemble alternative versions of the robot, which includes models resembling vehicles, machines, animals, and humanoids. In this chapter, we will work with a version of the LEGO NXT known as the TriBot. This is the simplest assembly, and the NXT kit includes a quick-start guide, which instructs you how to assemble the TriBot. You will need to go beyond the quick-start guide and assemble the bumper, sound control, and sonar sensors. Instructions for how to add these sensors are included with the documentation installed with your NXT. After you assemble it, your robot should resemble the robot in Figure 6-1.

> **Note** The bumper sensor depicted in Figure 6-1 is slightly different than the one described in the LEGO NXT assembly instructions. Instead of allowing the bumper to drag above the ground, the bumper is attached to the peg hole within the touch sensor. This allows the touch sensor to work more effectively as a bumper.

Figure 6-1 Assembled LEGO NXT, which is used in the sample code included with this chapter.

Working with Version 1

In this section, we will walk through the steps for creating a simple service named Wander, which is used to move the LEGO NXT around a room. Even though we will use generic services to operate the motors and contact sensors, we will include a reference to the LEGO NXT service. By including the LEGO service, we will be able to use the buttons on the *LEGO brick* to start moving the robot forward and also to stop the robot, if necessary. The downside to doing this is it will make the service specific to the LEGO NXT and not applicable to other types of robots. This is the tradeoff to consider when designing services that use other non-generic services.

> **LEGO brick** This represents the brains for the robot, and it is where the programmable chip for your NXT resides. This brick-shaped object includes ports for attaching the sensors and servo motors included with your NXT. It also includes four buttons that you can use to control the robot.

To begin you need to create a new Decentralized Software Services (DSS) service using the Visual Studio template installed with MSRS. To do this, open Visual Studio 2005 and create a new project named Wander using the Simple Dss Service (1.5) template. Doing so creates a new project that contains two source files: WanderTypes.cs, which contains classes that define the operations and state of the service, and Wander.cs, which is the implementation class.

> **Tip** When creating a new project, you will be asked to specify the project location. By default, Visual Studio points to a location within \documents\Visual Studio 2005\Projects. To run the Wander service, you need to execute the DssNode program as an external program and load two manifest files. One of these manifest files will be for the LEGO NXT robot, and this file resides within the MSRS application directory. For the application to locate this file properly, you need to save your Wander project to a location within the MSRS application directory. Instead of using the default path assigned, use the following path: C:\Microsoft Robotics Studio (1.5)\samples\Wander\WanderVersion1. C:\Microsoft Robotics Studio (1.5) is the default location where MSRS is installed. If you chose to install MSRS to a different location, then your path should reflect this.

Defining the Service Contract

The *WanderState* class is where you place declarations for the state variables used by your service. This class is located in the WanderTypes.cs file. You need to add code that declares the only state variable used by this service. The *RobotStatus* state variable will hold a string value that indicates what operation the robot is performing. For example, when the robot is driving forward, the state variable will contain a string that indicates the robot is driving. The code for this declaration is as follows:

```
// Add this code to create state variables that will indicate
// what the robot is currently doing
private string _robotStatus;

[DataMember]
public string RobotStatus
{
    get { return _robotStatus; }
    set { _robotStatus = value; }
}
```

The WanderTypes.cs file is also where you add interface code for any new service operations. You may recall from the last chapter that certain operations are implemented for you automatically when you create a new service using the Visual Studio template. For the Wander service, we need to add only one operation: Replace. We use this operation to replace the value of the only state variable used by the Wander service. To add the Replace operation, you first need to modify the existing PortSet. Locate the declaration for the main operations port in the *WanderTypes* class and change the class definition to look like the following:

```
[ServicePort()]
public class WanderOperations : PortSet<DsspDefaultLookup,
        DsspDefaultDrop,
        Get,
        Replace>
{
}
```

You also need to add the interface definition for the replace operation below the PortSet. The code for the replace operation is as follows:

```
// Replace the existing state
public class Replace : Replace<WanderState, PortSet<DefaultReplaceResponseType, Fault>>
{
}
```

Add Code to the Implementation Class

In this project, the implementation class is named *Wander.cs*. The implementation class is where the majority of your service code resides. It contains the code that reads incoming data from the touch and sonar sensors and sends commands to operate the robot's motors.

New LEGO NXT Services

In October 2007, the MSRS team released an update to the samples for MSRS that included a new set of services for the LEGO NXT. These services are known as version 2, and they work in parallel with the existing NXT services. This means that services written to work with the earlier version of the NXT services continue to work even if you install the new version 2 services.

The Wander services in this chapter are written to work with the new LEGO NXT services. To get these services, you must download them from the MSRS Web site (*http://www.microsoft.com/downloads*) and search the words "samples update robotics." If you used the original NXT services, you will find working with the new LEGO NXT services much easier.

Note Before downloading and installing the new LEGO NXT services, close all instances of Visual Studio 2005.

Note After you download and install the services, you need to ensure that your LEGO NXT is running the correct version of firmware. Do not assume that a recently purchased NXT has the latest firmware. Refer to the installation instructions included with the update, which, by default, are installed in the following location: C:\Microsoft Robotics Studio (1.5)\samples\SamplesUpdatePackage.htm. If you try to download the firmware to the NXT using the Bluetooth connection, it will not be able to locate the device.

Set References

Before you begin adding code to the implementation class, you will need to add references to the project. To do this, right-click the References folder in Solution Explorer and click Add Reference. From the Add Reference dialog box, scroll through the components listed in the .NET tab and locate the following (after you are done adding the references, Solution Explorer should appear similar to Figure 6-2):

- **Nxtbrick.y2007.m07.proxy** This assembly provides access to the supported functions specific to the LEGO NXT. This includes control over the robot's sensors and motors. For the Wander service, you need to reference the *ButtonsOperations* class in order to subscribe to the buttons on the LEGO brick.
- **NxtCommon.Y2007.M07** This assembly (which, in this case, is NOT the proxy) contains reusable LEGO NXT data contracts and methods.
- **Robotics.Common.Proxy** This assembly, which is included with MSRS, allows you to access a variety of namespaces used to control a robot. In this chapter, we will add a reference to the *Microsoft.Robotics.Services.Drive.Proxy* namespace. This gives you access to drive operations used to operate a two-wheel differential drive system.

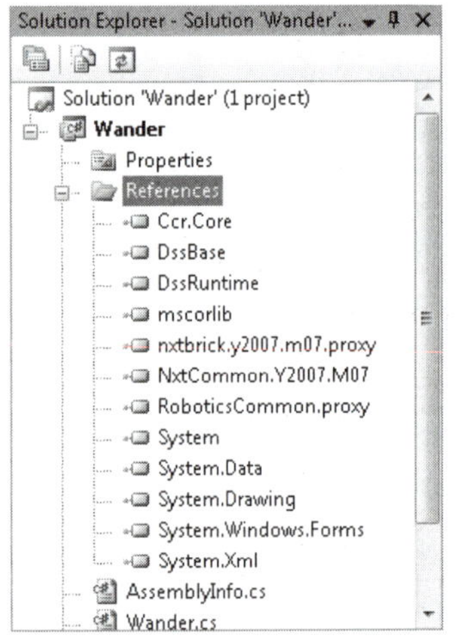

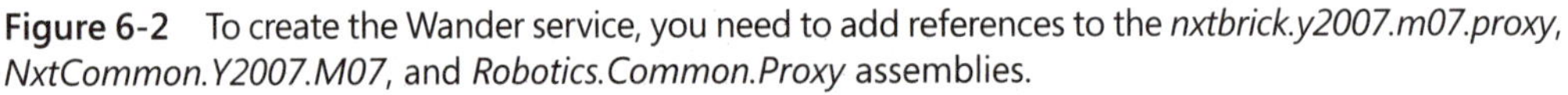

Figure 6-2 To create the Wander service, you need to add references to the *nxtbrick.y2007.m07.proxy*, *NxtCommon.Y2007.M07*, and *Robotics.Common.Proxy* assemblies.

After you add the references, you need to add code for namespace declarations. The following code can be added to the top of the Wander.cs file:

```
using W3C.Soap; //Gives us access to the Fault object
using drive = Microsoft.Robotics.Services.Drive.Proxy;

// Add references to the LEGO services
using lego = Microsoft.Robotics.Services.Sample.Lego.Nxt.Buttons.Proxy;
using bumper = Microsoft.Robotics.Services.Sample.Lego.Nxt.TouchSensor.Proxy;
```

Add Partnerships

Just like the BasicDrive service from the last chapter, the Wander service uses the generic drive service, which is part of the *Robotics.Common.Proxy* assembly. To use this service, you must first declare a partnership using the *Partner* attribute. You also need to add partnerships that point to the *ButtonOperations* class and *TouchSensorOperations* class. You should place the code to add these partnerships and declare the notification ports associated with them below the declaration for the main port, as follows:

```
// Partner: NxtTouchSensor,
[Partner("NxtTouchSensor", Contract = bumper.Contract.Identifier,
          CreationPolicy = PartnerCreationPolicy.UsePartnerListEntry)]
bumper.TouchSensorOperations _nxtTouchSensorPort =
           new bumper.TouchSensorOperations();
bumper.TouchSensorOperations _nxtTouchSensorNotify =
           new bumper.TouchSensorOperations();

// Add a partner to the NXT drive service
[Partner("NxtDrive", Contract = drive.Contract.Identifier,
          CreationPolicy = PartnerCreationPolicy.UseExisting)]
private drive.DriveOperations _drivePort = new drive.DriveOperations();

// Add partner for the LEGO NXT buttons to start the forward movement
[Partner("buttons", Contract = lego.Contract.Identifier,
          CreationPolicy = PartnerCreationPolicy.UseExisting)]
private lego.ButtonOperations _legoPort = new lego.ButtonOperations();
private lego.ButtonOperations _legoNotifyPort =
          new lego.ButtonOperations();
```

Modify the *Start* Method

When you create a service using the Visual Studio template, a method named *Start* is created automatically. This method is called when the service starts, and it is where you add initialization code specific to your service. For the Wander service, you need to replace the code in the existing *Start* method with the following:

```
protected override void Start()
{
    //Check to see if the state already exists.
    //If not, then we will initialize it
    if (_state == null)
    {
       _state = new WanderState();
       _state.RobotStatus = "Waiting for command";
    }
```

```
    base.Start();

    // Define our Notification handlers using the CCR Interleave primitive
    Activate(
      Arbiter.Interleave(
        new TeardownReceiverGroup(),
        new ExclusiveReceiverGroup(),
        new ConcurrentReceiverGroup(
            Arbiter.Receive<bumper.TouchSensorUpdate>(true,
              _nxtTouchSensorNotify, TouchSensorHandler),
            Arbiter.Receive<lego.ButtonsUpdate>(true,
              _legoNotifyPort, LegoHandler)
              )
            )
        );

    // Subscribe to the notification ports. This tells us whether
    // the bumper has been pressed or the button on the LEGO brick
    // was pressed
    nxtTouchSensorPort.Subscribe(_nxtTouchSensorNotify);
    legoPort.Subscribe(_legoNotifyPort);
}
private void TouchSensorHandler(bumper.TouchSensorUpdate notification){}
private void LegoHandler(lego.ButtonsUpdate notification) {}
```

Note This code block also contains method stubs to represent the message handlers. For now the body of these handlers is empty; we cover what code goes in these handlers in the next section.

In the preceding code, we use a Concurrency and Coordination Runtime (CCR) primitive known as the Interleave. The Interleave arbiter allows us to group multiple notification handlers according to how they should be executed. In this case, we are assigning all calls to the same group, which is known as the ConcurrentReceiverGroup. This group specifies that the code within it will run in parallel with each other. One thing to keep in mind is that the ConcurrentReceiverGroup does not have precedence over code that runs with an Update or Stop operation. Alternatively, if we would have used the ExclusiveReceiverGroup, then the code is guaranteed to run while no other handler is running. Code specified in the TeardownReceiverGroup executes only one time, and no other messages are processed.

After the notification handlers are declared, we need to subscribe to the notification ports defined in our partnership declarations. For the Wander service, we will receive notifications from ports associated with the generic drive, generic contact sensors, and LEGO services.

Create Message Handlers

Now it is time to complete the message handlers that were referenced in the *Start()* method. The first handler to complete is the *LegoHandler*. This handler is called every time the LEGO brick encounters a change. For the Wander service, we will use the left and right buttons on the LEGO brick (see Figure 6-3). Pressing the left button will instruct the robot to move forward, and pressing the right button will instruct the robot to stop.

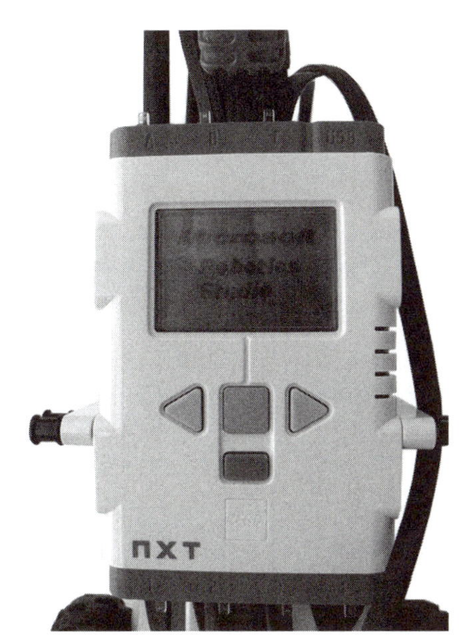

Figure 6-3 Close-up of the LEGO NXT brick. This shows the left and right buttons, which are used to trigger behavior in the sample code provided with this chapter.

The code for the *LegoHandler* is as follows:

```
private void LegoHandler(lego.ButtonsUpdate notification)
{
      // If the left button was pressed, move the robot forward
      if (notification.Body.PressedLeft)
      {
         SpawnIterator<double, double>(0.75, 0.75, DriveRobot);
         LogInfo("left button was pressed");
         return;
      }
      // If the right button was pressed, stop the robot
      else if (notification.Body.PressedRight)
      {
          SpawnIterator<double, double>(0, 0, DriveRobot);
          LogInfo("right button was pressed");
          return;
      }
}
private IEnumerator<ITask>DriveRobot(double leftWheelPower, double
                rightWheelPower)
{
    yield return;
}
```

> **Note** This code block contains a method stub that represents the *DriveRobot* function. We cover the body for this function later in the section.

LegoHandler receives a notification message from the *ButtonsUpdate* class, which is part of *ButtonOperations*. The *ButtonsUpdate* class indicates any change to the state of the LEGO brick, and it includes members to represent each of the LEGO buttons (Left, Right, Cancel, and Enter). We simply need to check the value for the *PressedLeft* property. If it is set with a value of True, then we know the left button on the brick has been pressed, and we can instruct the robot to start moving forward. If the *PressedRight* property is set with a value of True, then we know the right button was pressed, and we want to tell the robot to stop.

The *LegoButtons* handler uses a function named *DriveRobot* to give the robot instructions. *SpawnIterator* is a CCR function that allows you to invoke an iterator-based message handler asynchronously. This means that the code executed within the *DriveRobot* function is managed by the CCR, and completing its execution does not slow anything else down. The *DriveRobot* function, which you see below, accepts two input parameters; one that specifies the amount of power for the left wheel and one that specifies the amount of power for the right wheel. You need to insert the code for the *DriveRobot* function. We added the method stub for this function in an earlier step.

```
private IEnumerator<ITask>DriveRobot(double leftWheelPower, double
                rightWheelPower)
{

    // Create a request that will start driving the robot based
    // on the left and right power settings submitted
    drive.SetDrivePowerRequest driveRequest = new
               drive.SetDrivePowerRequest();
    driveRequest.LeftWheelPower = (double)leftWheelPower;
    driveRequest.RightWheelPower = (double)rightWheelPower;

    // Send the request to the main port
    yield return Arbiter.Choice(
         _drivePort.SetDrivePower(driveRequest),
         delegate(DefaultUpdateResponseType response) { },
         delegate(Fault fault)
         {
            LogError(null, "Unable to drive robot", fault);
         }
     );
     //Update the robot status
     state.RobotStatus = "Robot is driving";
}
```

The *DriveRobot* function creates a request named *SetDrivePowerRequest*. This special request type is used whenever there is a need to set the drive power using the generic drive service. The SetDrivePowerRequest lets you specify the power setting for both the left and right wheels.

The *LeftWheelPower* and *RightWheelPower* properties can be set with values ranging from −1.0 to 1.0. This value indicates the percentage of power that is applied to the motors that power each wheel. If you use the same positive value for each wheel, then the robot moves forward. Alternatively, if you use the same negative values for each wheel, then the robot moves backward. You can also use alternating values, which cause the robot to turn in a particular direction.

After the request is created, you can send it to the drive port using the *Arbiter.Choice* method. If the request returns an error, the following message is written to the log: "Unable to drive robot."

The last handler to add code for is the one that is triggered by an update to the NXT touch sensor. The code that should be inserted into the *TouchSensorHandler* is as follows:

```
private void TouchSensorHandler(bumper.TouchSensorUpdate notification)
{
     //See if the touch sensor has been pressed
     if (notification.Body.TouchSensorOn)
     {
         SpawnIterator(ReverseAndTurn);
         LogInfo("touch sensor bumper was pressed");
     }
}
```

The *TouchSensorOn* property returns a Boolean value, which indicates whether the touch sensor has been triggered. After you determine that the sensor was pressed, you can invoke the *ReverseAndTurn* method using *SpawnIterator*. The code for this method, which you should insert below the *TouchSensorHandler* method, is as follows:

```
private IEnumerator<ITask> ReverseAndTurn()
{

     // First request that the robot reverse
     SpawnIterator<double, double>(5, -.50, DriveRobotSetDistance);

     // Wait for 1/2 second or 500 milliseconds to prevent
     // the drive commands from interfering with each other
     yield return Arbiter.Receive(false, TimeoutPort(500),
        delegate(DateTime t) { });

     // Turn to the right
     SpawnIterator<double, double>(90, 0.50, TurnRobot);

     // Wait for 1/2 second or 500 milliseconds to prevent
     // the drive commands from interfering with each other
     yield return Arbiter.Receive(false, TimeoutPort(500),
       delegate(DateTime t) { });

     // Go forward again
     SpawnIterator<double, double>(0.75, 0.75, DriveRobot);
```

```
        // terminate the iteration
        yield break;
    }
    private IEnumerator<ITask> DriveRobotSetDistance(double distance,
           double power) {}
    private IEnumerator<ITask> TurnRobot(double angle, double power){}
```

> **Note** This code block contains method stubs that represent the *DriveRobotSetDistance* and *TurnRobot* functions. The body for these functions are covered later in the section.

ReverseAndTurn is called whenever the robot hits an obstacle, so it follows that the code would initiate a request to back up and turn the robot away from the obstacle. This is done by invoking a series of calls to drive functions used to perform specific tasks. For example, the *DriveRobotSetDistance* function can be used to reverse the robot for a set distance. It does this by using the *DriveDistanceRequest* and *DriveDistance* functions, which are provided with the generic drive service. The code for the *DriveRobotSetDistance* function is shown as follows (you need to add code for the function only because we created the function stub in an earlier step):

```
private IEnumerator<ITask> DriveRobotSetDistance(double distance,
        double power)
{

    // Create a request to drive a specific distance
    drive.DriveDistanceRequest distanceRequest = new
          drive.DriveDistanceRequest();
    // millimeters to meters
    distanceRequest.Distance = (double)distance / 1000.0;
    distanceRequest.Power = (double)power;

    // Send the request to the drive port
    yield return Arbiter.Choice(
       _drivePort.DriveDistance(distanceRequest),
        delegate(DefaultUpdateResponseType response) { },
        delegate(Fault fault)
       {
         LogError(null, "Unable to drive robot specified distance", fault);
       }
    );

    //Update the robot status
    state.RobotStatus = "Robot is driving a set distance";

}
```

By using this method, we can ensure that the robot reverses its direction for a specific distance. The alternative is to call the *DriveMotor* function using reverse power levels and then use a timer to stop that movement after so many milliseconds elapses. This is a less precise method than the *DriveRobotSetDistance* function, so it is therefore less preferable.

We still use a timer to control how often drive commands are sent to the robot. This ensures that the robot is not bombarded with multiple commands, which might cause it to throw an error. The *TimeoutPort*, which is part of the CCR base, can be used to pause execution for a specified time span. The time span, which is in milliseconds, specifies the amount of time that must pass before a timeout occurs and a value is returned. The fact that we use the *yield* keyword means that the processing should wait until the timeout has occurred.

A turn to the right is initiated by a call to the *TurnRobot* function. The *TurnRobot* function utilizes the *RotateDegreesRequest* and the *RotateDegrees* functions, which are provided with the generic drive service. The *RotateDegrees* function allows you to specify an angle in which the robot must turn. The code for the *TurnRobot* function is shown as follows (you need to add code for the function only because we created the function stub in an earlier step):

```
private IEnumerator<ITask> TurnRobot(double angle, double power)
{

     // Create request to drive robot according to a degree of rotation
     drive.RotateDegreesRequest turnRequest =
          new drive.RotateDegreesRequest();
     turnRequest.Degrees = (double)angle;
     turnRequest.Power = (double)power;

     // Send the request to the drive port
     yield return Arbiter.Choice(
     drivePort.RotateDegrees(turnRequest),
        delegate(DefaultUpdateResponseType response) { },
        delegate(Fault fault)
        {
            LogError(null, "Unable to turn the robot", fault);
        }
      );

      //Update the robot status
      state.RobotStatus = "Robot is turning";
}
```

The alternative to calling the *TurnRobot* function is to use the *DriveMotor* function to send a positive power level to the left wheel motor, while also sending a zero power level to the right motor. This would cause the left wheel to spin and the right wheel to stay still. This method is less precise than the method used in *TurnRobot* and, thus, less preferable. After calling the *TurnRobot* function, we also use a yield return to the *TimeoutPort*, which causes a delay of 500 milliseconds.

Change the Wander Manifest

As with the service in the last chapter, you need to make changes to the Wander manifest before you can run version 1. Even though you can edit the XML-based manifest file, which is named Wander.manifest.xml, using a text-based editor such as Windows Notepad, it is best to use the DSS Manifest Editor provided with MSRS. You can also use this editor to adjust configuration settings for the service.

Before you can change the manifest, you need to successfully compile the service because this generates assemblies used by the Manifest Editor. To do this, select Build, and then click Build Solution. Ensure that the status in the bottom-left status bar shows that the build succeeded.

Associate Services

When you open the DSS Manifest Editor, you see a list of services on the left. Scroll through this list until you locate the Wander service. This is created when you compiled your project successfully. Drag an instance of the Wander service onto the design surface. This shows three icons, which represent each service defined as a partner. The Manifest Editor shows an arrow next to the icon and a box, which declares that the manifest will use the service's partner definition (see Figure 6-4).

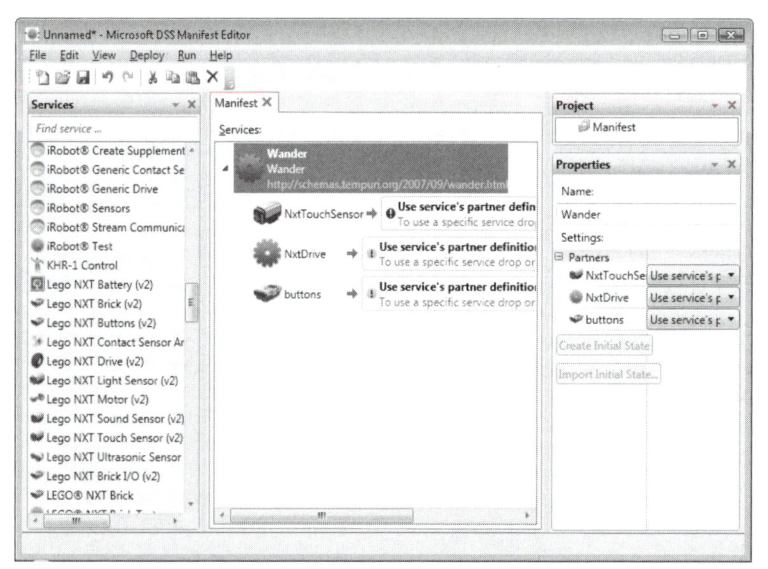

Figure 6-4 The first step in building the manifest for the Wander service is to drag an instance of the Wander service onto the design surface.

To help our service run more efficiently, we can associate the partners with a specific LEGO-based service. We do this by dragging an instance of the hardware-specific service onto the box in the design area. For example, we can drag an instance of the LEGO NXT Drive (v2) service onto the box next to the partner named NxtDrive. Additionally, we can drag an instance of the LEGO Touch Sensor (v2) onto the NxtTouchSensor and the LEGO NXT Buttons (v2) onto the buttons. After it's completed, the manifest should resemble Figure 6-5.

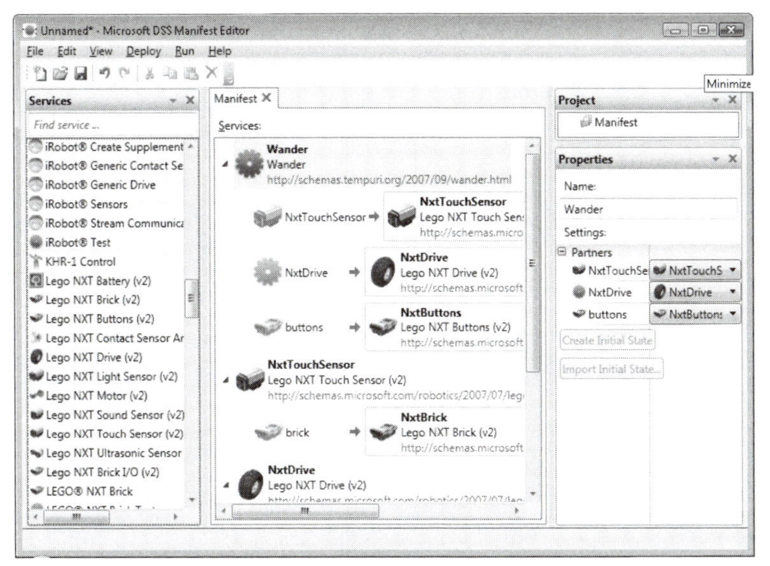

Figure 6-5 Completed manifest for the Wander service, version 1.

Tip When you click on the partner box in the design surface, the service list on the left shows only the services that can be associated with that generic service.

Configure Services

You can also use the Manifest Editor to configure your services. To do this, select the service, click Edit, and then click Set Configuration. This displays properties in the right-hand pane. From here you click Create Initial State, and additional state properties appear in the pane. Each service has different state properties exposed, and not all of the properties need to be configured (see Figure 6-6).

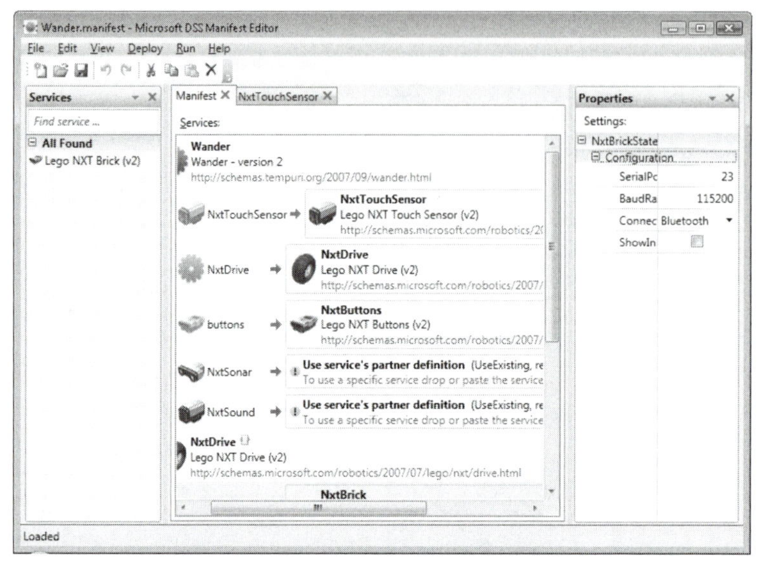

Figure 6-6 The Properties pane in the DSS Manifest Editor allows you to configure service state for the Wander service.

Table 6-1 lists the state properties that you need to configure for the Wander service.

Table 6-1 Configuration Properties for the Wander Service

Partner Service	State Property	Value
NxtTouchSensor	Brick	Select NxtBrick
	Name	Sensor1
	SensorPort	Sensor1
NxtDrive	Brick	Select NxtBrick
	DistanceBetweenWheels	0.122
	LeftWheel/MotorPort	MotorC
	LeftWheel/WheelDiameter	0.055
	RightWheel/MotorPort	MotorB
	RightWheel/WheelDiameter	0.055
NxtButtons	Brick	Select NxtBrick
Brick	Configuration/SerialPort	Number of the Outgoing Port assignment for your Bluetooth connection

The final step is to save a copy of the manifest. If you click File and Save As, you are prompted to select a file path. Browse to the location where the Wander service is located and select the Wander.manifest file. Select yes when asked whether you want to replace the existing file.

> **Note** When you save the manifest file, you will notice that additional config files are added to the directory where you saved the manifest file. These config files are created by the Manifest Editor, and they hold the configuration values you just set.

Change Project Properties

The final step is to ensure that the services for the LEGO NXT are loaded with the Wander service. This is done by including a LEGO manifest file when calling the DssHost program. You may recall from Chapter 2, "Understanding Services," that the DssHost program is a command-line tool used to start a DSS node on a port. This is needed in order to run a service. When calling this command-line tool, you specify one or more manifest files that will be included in the startup.

When you create a service using the Visual Studio template, it automatically adds a call to the DssHost program as part of the project's debug properties. You can see these properties by clicking Project and then selecting Wander Properties. The Debug tab displays Start options that include command-line arguments. By default, the manifest file for the Wander service is appended to the argument list. You also need to add a reference to a LEGO NXT manifest.

MSRS provides manifest files for supported robots in the samples\config folder for your local MSRS installation. There are several manifest files for the LEGO NXT. Which one is needed depends on the kinds of functions the LEGO NXT needs to perform. For the Wander service, we need to use the LEGO.NXT.MotorTouchSensor.manifest because we will be accessing services that control the motors and touch sensors. To add this manifest file, place your cursor at the end of the text inside the command-line arguments text box. Add a single space, and then add the following text:

```
-m:"samples\config\LEGO.NXT.MotorTouchSensor.manifest.xml"
```

After you make these changes, you need to save the project by clicking File and then clicking Save All. You are now ready to run the Wander service and find out how your robot will perform.

Run the Wander Service

Before running the Wander service, you need to ensure that your LEGO NXT is assembled and configured to work with your development machine. Keep in mind that you must use Bluetooth to connect to your NXT. This is because the interface programs provided with MSRS allow only for a Bluetooth connection. This means your development machine needs to be Bluetooth enabled, or you need to use a supported Bluetooth USB dongle. Refer to the section titled "Configuring the LEGO Mindstorms NXT" in Appendix B for instructions on how properly to configure the LEGO NXT.

To run the Wander service from Visual Studio, you need to press Ctrl+F5 to start without debugging or press F5 to start with debugging. Either way, you should see a command-line window appear and, within this window, text that indicates the service is being loaded. After it loads successfully, you should hear a beep from the LEGO NXT, and the light sensor should turn on. At this point, you can press the left button on the LEGO brick to start moving the robot forward. You can either push the bumper directly or allow the robot to run into an object that triggers the bumper sensor. When the bumper is pressed, the robot should back up, turn to the right, and then start moving forward again.

Evaluate the Robot's Behavior

If you have followed along with the chapter, or if you are using a copy of the version 1 Wander service from the book's companion Web site, then you will see very quickly what problems this version of the Wander service has. For starters, a collision has to occur for the LEGO NXT to move in a different direction.

There are two major problems with using this type of reactive approach. The first problem with the reactive approach is that you are taking a risk that the robot might be damaged. In many cases, robotics hardware is very expensive and not easily replaced. If the robot runs off a high ledge or down a flight of stairs, it could easily be damaged beyond repair. The second problem with this approach is that the bumper sensor will not always trigger a collision. If you take a look at the assembled robot in Figure 6-1, you will see that the touch sensor is triggered by a small T-shaped rod directly attached to the NXT's touch sensor. It is possible for the robot's wheels to get hung up on a low-lying obstacle and for the bumper sensor never to be triggered. In this state, the robot could sit there spinning its wheels indefinitely.

Try experimenting with version 1 of the Wander service and see what other potential problems you can see with the code. In the next section, we will make a few changes to the Wander service that enables it to overcome some of these problems.

Working with Version 2

Version 2 of the Wander service corrects a few problems you may have encountered while running version 1 of the service. Before making changes to the code you have created so far, you should make a copy of the existing Wander service and name the source folder Version1. You can then name the folder containing the copied version Version2.

In version 2, we will integrate the ultrasonic/sonar sensor available on the LEGO NXT. This will allow the NXT to turn away from an obstacle before a collision occurs. We will also use the sound sensor to detect a loud noise and, thus, command the robot to stop. Finally, we will add a watchdog timer to help prevent the robot from "spinning its wheels" unnecessarily.

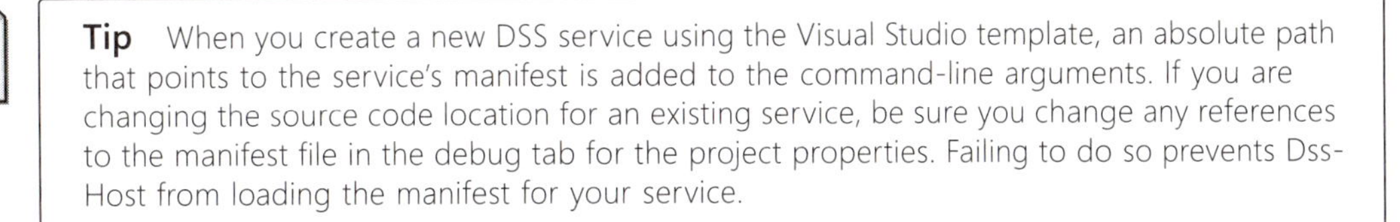

Tip When you create a new DSS service using the Visual Studio template, an absolute path that points to the service's manifest is added to the command-line arguments. If you are changing the source code location for an existing service, be sure you change any references to the manifest file in the debug tab for the project properties. Failing to do so prevents Dss-Host from loading the manifest for your service.

Add the Sonar Sensor

The first thing to add to the Wander service is a reference to the NXT *Sonar* namespace. You should add the following code below the other references in the Wander service:

```
using sonar = Microsoft.Robotics.Services.Sample.Lego.Nxt.SonarSensor.Proxy;
```

The next thing to add is a partner attribute, which represents the NXT sonar service. This is used to subscribe to the NXT's sonar sensor, and the code should be added below the other partner declarations.

```
[Partner("sonar", Contract = sonar.Contract.Identifier,
       CreationPolicy = PartnerCreationPolicy.UseExisting)]
private sonar.UltrasonicSensorOperations _sonarPort = new
       sonar.UltrasonicSensorOperations();
private sonar.UltrasonicSensorOperations _sonarNotifyPort = new
       sonar.UltrasonicSensorOperations();
```

You need to add a line to the *ConcurrentReceiverGroup*, which registers the handler for the sonar sensor. Add the following line of code to the Interleave pattern located in the *Start* method:

```
Arbiter.Receive<sonar.SonarSensorUpdate>(true, _sonarNotifyPort, SonarHandler),
```

Additionally, you need to subscribe to the sonar notification port, which in this case is named _sonarNotifyPort. This is accomplished by adding the following code to the *Start* method:

```
_sonarPort.Subscribe(_sonarNotifyPort);
```

You now need to create the handler that is invoked each time a notification arrives on the sonar notification port. The following code can be used for the *SonarHandler*, which can be added below the other message handlers:

```
private void SonarHandler(sonar.SonarSensorUpdate notification)
{
    // Call a function to get the most recent notification since
    // we may have several of them queued up before this handler
    // was called
    sonar.SonarSensorState sonarData =
      GetMostRecentSonarNotification(notification.Body);

    // check to see that we have some valid data to process
    if (sonarData.Distance != 0)
```

```
    {
        if (sonarData.Distance < 30)
        {
            SpawnIterator(ReverseAndTurn);
            LogInfo("the sonar detected an object too close to the robot");
           _state.MostRecentNotification = sonarData.TimeStamp;
        }
    }
}
private sonar.SonarSensorState GetMostRecentSonarNotification(sonar.SonarSensorState
sonarData)
{ Return sonarData; }
```

Note This code block contains method stubs that represent the *GetMostRecentSonarNotification* function. The body for this function will be covered later in the section.

The *SonarHandler* makes a call to the *GetMostRecentSonarNotification* function, which passes back the message for the most recent notification. It is possible for the sonar sensor to record multiple readings before the sonar handler has been called. In these cases, we deal only with the last notification. The code for the *GetMostRecentSonarNotification* is as follows (you need to add the code only to the method stub that we created in an earlier step):

```
private sonar.SonarSensorState GetMostRecentSonarNotification(sonar.SonarSensorState
sonarData)
{
    sonar.SonarSensorUpdate temp;

    // Get the number of messages queued up on the sonar notification port
    int count = _sonarNotifyPort.Ports.Count - 1;

    // Loop through all the messages queued up on this port
    // since we need to find the last one
    for (int i = 0; i < count; i++)
    {
        // The test method will remove a message from the port
        temp = _sonarNotifyPort.Test<sonar.SonarSensorUpdate>();

        // Return the last one by comparing the timestamps
        if (temp.Body.TimeStamp > sonarData.TimeStamp)
        {
            sonarData = temp.Body;
        }
    }

    // Send back the last notification message
    return sonarData;
}
```

The *Distance* property contains the distance in centimeters to the nearest detectable object. If this value is below a certain threshold value (that you define), then an object is too near and a collision is imminent. Therefore, you want to command the robot to reverse and turn away from the object.

Add the Sound Sensor

In version 1, the only way you could stop the robot was to push a button on the LEGO brick. Additionally, we can use the sound sensor to provide an alternative shutdown mechanism. Both the sound and light sensors on the LEGO NXT are known as *analog sensors*. You could use the generic analog sensors service to access this sensor, but for the LEGO NXT, a LEGO NXT Sound Sensor is specifically provided. Use the following code to add a reference to this namespace:

```
using sound = Microsoft.Robotics.Services.Sample.Lego.Nxt.SoundSensor.Proxy;
```

Analog Sensor Analog sensors are devices used to measure signals of varying strength, such as an audio signal. Examples of analog sensors include sensors that measure sound, light, and temperature.

You will also need to add a partner declaration for the LEGO NXT Sound Sensor service. The following code can be placed below the other partner declarations:

```
// Add partner for the NXT sound
[Partner("NxtSound", Contract = sound.Contract.Identifier,
    CreationPolicy = PartnerCreationPolicy.UseExisting)]
private sound.SoundSensorOperations _soundPort = new
    sound.SoundSensorOperations();
private sound.SoundSensorOperations _soundNotifyPort = new
    sound.SoundSensorOperations();
```

As with the sonar sensor, you will need to add the following code to the *Start* method, which registers the handler named *SoundHandler*.

```
Arbiter.Receive<sound.SoundSensorUpdate>(true, _soundNotifyPort, SoundHandler)
```

Additionally, you need to add the following code to the *Start* method to subscribe to the service and receive notifications from the sound notify port:

```
_soundPort.Subscribe(_soundNotifyPort);
```

The next thing to add is the following code for the handler named *SoundHandler*:

```
private void SoundHandler(sound.SoundSensorUpdate notification)
{

    // check to see that the sound is above
    // a 35% sensor reading level
    if (notification.Body.Intensity > 35)
    {
        // Stop the robot
```

```
            SpawnIterator<double, double>(0, 0, DriveRobot);
            LogInfo("the sound sensor detected a noise; robot will now stop");
            state.MostRecentNotification = notification.Body.TimeStamp;
        }

}
```

If the *Intensity* property contains a value greater than 35, it commands the robot to stop.The LEGO NXT sound sensor detects decibels and returns the data in the form of a percentage. The higher the percentage, the louder the sound. In the *SoundHandler* for version 2, we are using a threshold value of 0.35, or 35 percent. You may have to experiment with your LEGO NXT to find the threshold level that works properly for your environment.

Add a Watchdog Timer

Because it is possible for the wheels of the LEGO NXT to get caught on a surface or for sensors to stop functioning correctly, it is a good idea to have a watchdog timer added to your roaming service. The watchdog timer acts as a safety guard that stops the robot if no sensor activity is reported over a period of time. To add a watchdog timer, we need to add a state variable that holds the data and time the last sensor notification was received. To create this state variable, you should add the following code to the *WanderTypes* class file:

```
private DateTime _mostRecentNotification;

[DataMember]
public DateTime MostRecentNotification
{
    get { return _mostRecentNotification; }
    set { _mostRecentNotification = value; }
}
```

You also need to add an operation named *WatchDogUpdate* to the PortSet for the *Wander-Operations*, along with the following code:

```
public class WatchDogUpdate : Update<WatchDogUpdateRequest,
PortSet<DefaultUpdateResponseType, Fault>>
{
    public WatchDogUpdate(WatchDogUpdateRequest body) : base(body) { }

    public WatchDogUpdate() { }
}

[DataContract]
public class WatchDogUpdateRequest
{
    private DateTime _timeStamp;

    [DataMember]
    public DateTime TimeStamp
    {
        get { return _timeStamp; }
```

```
        set { _timeStamp = value; }
    }

    public WatchDogUpdateRequest(DateTime timeStamp)
    {
        TimeStamp = timeStamp;
    }

    public WatchDogUpdateRequest()
    { }
}
```

Because the watchdog timer will be updating the service state, you need to add the handler to the exclusive receiver group. Add the following code to the *Start* method in the *Wander* class file:

```
Arbiter.Receive<WatchDogUpdate>(true, _mainPort, WatchDogUpdateHandler)
```

After adding the preceding statement, the section of code that declares all the handlers should look like the following:

```
Activate(
     Arbiter.Interleave(
       new TeardownReceiverGroup(),
       new ExclusiveReceiverGroup(
          Arbiter.Receive<WatchDogUpdate>(true, _mainPort,
             WatchDogUpdateHandler)),
       new ConcurrentReceiverGroup(
          Arbiter.Receive<bumper.TouchSensorUpdate>(true,
           _nxtTouchSensorNotify, TouchSensorHandler),
          Arbiter.Receive<lego.ButtonsUpdate>(true,
           _legoNotifyPort, LegoHandler),
          Arbiter.Receive<sonar.SonarSensorUpdate>(true,
           _sonarNotifyPort, SonarHandler),
          Arbiter.Receive<sound.SoundSensorUpdate>(true,
           _soundNotifyPort, SoundHandler)
          )
     )
);
```

We also need to add code that starts the watchdog timer by posting an update to the main port. Before this happens, we need to initialize the watchdog state variable with the current date and time. The following code can be added below the current subscriptions in the *Start* method:

```
// Initialize the WatchDog State with the current time
// and start the timer
_state.MostRecentNotification = DateTime.Now;
_mainPort.Post(new WatchDogUpdate(new
        WatchDogUpdateRequest(DateTime.Now)));
```

The final thing to add is the handler for the watchdog update. This is named *WatchDogUpdate-Handler* and is represented in the following code:

```
private void WatchDogUpdateHandler(WatchDogUpdate update)
{
    // Get the time span since the last notification from a sensor
    TimeSpan sinceNotification = update.Body.TimeStamp -
            _state.MostRecentNotification;

    // See if the latest notification was received over 30 seconds ago
    if (sinceNotification.TotalSeconds >= 30)
    {
        // Stop the robot
        SpawnIterator<double, double>(0, 0, DriveRobot);
        LogInfo("Last sensor reading: " + _state.MostRecentNotification);
    }

    // Start back the timer to fire every 500 milliseconds
    Activate(
       Arbiter.Receive(false, TimeoutPort(500), delegate(DateTime ts)
          {
           _mainPort.Post(new WatchDogUpdate(
               new WatchDogUpdateRequest(ts)));
           }
        )
    );

    update.ResponsePort.Post(DefaultUpdateResponseType.Instance);
}
```

The *WatchDogUpdateHandler* first computes the time span since the last sensor notification. If this value is greater than 30 seconds, it stops the robot. Regardless, it reactivates the watchdog update by using a TimeoutPort set to expire in 500 milliseconds. This represents the amount of time until the *WatchDogUpdateHandler* is called again.

Change the Wander Manifest

Just like with version 1, we need to alter the manifest for this version. Once again, it is best to use the DSS Manifest Editor to do this. Remember, you will have to first successfully compile the new version of the Wander service before opening the DSS Manifest Editor. After you do this, you can locate the Wander service in the list of services and drag it onto the design surface. You should see entries for the following partners: NxtTouchSensor, NxtDrive, NxtButtons, NxtUltrasonicSensor, and NxtSoundSensor. You need to drag an NXT-specific service onto the box located next to each of these partners.

The process is similar to what was done for the last version. This time you have two more partners to deal with: the sonar and sound sensors. When complete, the Manifest Editor should look like Figure 6-7. Make sure you save a copy of the manifest and replace the one that already exists in the application directory for Wander service version 2.

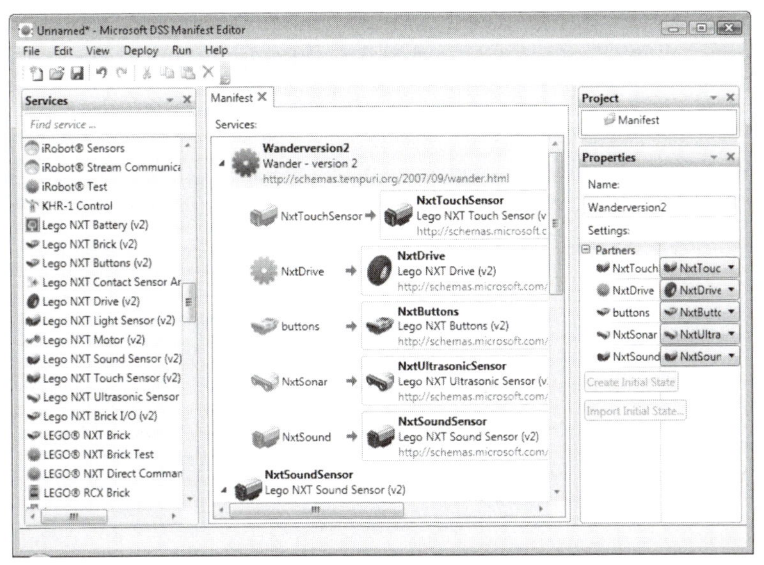

Figure 6-7 The manifest for version 2 of the Wander service, as it appears in the DSS Manifest Editor.

Evaluate the Robot's Behavior

If you run version 2 of the Wander service, you should notice some immediate differences. Most importantly, the robot should not collide with an object before it turns to move away from the obstacle. Try experimenting with different obstacles to get an idea of how sensitive the NXT sonar sensor is. You may have to adjust the threshold value used in the *SonarHandler*. Keep in mind that not every object will be detected by the sonar sensor, so the bumper sensor might still have to be used.

By incorporating the sound sensor and watchdog update timer, you have tools in place that can stop your robot in case of an emergency. A loud clapping noise should trigger the sound sensor and, in case the wheels get stuck, the robot should shut down after 30 seconds if there is no feedback from the sensors.

Summary

- This chapter walks through the steps for creating a service that allows a LEGO NXT to wander autonomously. The built-in sensors (touch, sound, and sonar) on the LEGO NXT will be used to operate the robot without remote control.
- Version 1 of the Wander service uses the NXT's touch sensor to act as a bumper. When it is triggered, the robot is commanded to back up and turn away from the object before moving forward again.

- Before running the Wander service, you will need to use the DSS Manifest Editor to specify NXT-based services that correlate with the NXT partners defined in the service. You will also need to modify the properties for the project so that the LEGO NXT manifest is loaded along with the service manifest.
- Version 2 of the Wander service expands upon the first and includes the use of the sonar sensor. This sensor is used to detect obstacles near the robot and command it to turn away before colliding with the object. Version 2 also uses the sound sensor as a signal to stop the robot, as well as a watchdog timer that times out after 30 seconds of no sensor activity.

Chapter 7

Creating a New Hardware Interface

Until now, this book has focused on robotics hardware supported by Microsoft Robotics Studio (MSRS). This means that the MSRS installation contains the onboard interface and service files necessary to interface with and operate a particular robot. But what if you want to work with your own custom-built robot or a robot kit not already supported by MSRS?

To operate an unsupported robot, you need to create an onboard interface that executes directly on the robotics hardware. You also must create services that not only communicate with the onboard interface but are also used to perform specific functions such as subscribe to a robot's contact sensors or send commands to drive the robot. In this chapter, we will review the steps needed to operate an unsupported robot named ARobot. Even if you decide not to purchase and assemble the ARobot, the sections in this chapter provide a good understanding of the steps necessary to build your own hardware interface with MSRS.

Working with the ARobot

ARobot is a three-wheeled robot kit (see Figure 7-1) made by a small robotics company in Texas named Arrick Robotics. ARobot was designed and built by the company's owner, Roger Arrick (see the sidebar titled "Profile: Roger Arrick"). The ARobot, pronounced "a robot," was created for robot hobbyists, and, like the BoeBot, it uses a Basic Stamp 2 (BS2) controller chip to provide the brains for the robot.

Figure 7-1 The ARobot by Arrick Robotics is an expandable, three-wheeled robot designed for robotics hobbyists.

Profile: Roger Arrick

Roger Arrick, Owner of Arrick Robotics

Roger Arrick is the owner of Arrick Robotics (*www.robotics.com*), a small electronics company based in Tyler, Texas. After moving out of the increasingly unprofitable computer accessory business, Arrick used his interest and excitement for robotics to venture into the unknown. He started the company in 1987 with no college degree, $1,000, and what he describes as "a wheelbarrow of stubbornness."

When asked about why he loves robotics, Arrick responded with:

> *Ever since I was knee-high to a solenoid, taking things apart, learning how they worked, and building things have been part of my inner being. Starting with Lincoln Logs, progressing to motorized Erector sets, then graduating to computers, I could not get enough. While my friends played sports and watched TV, I salvaged speakers from old radios and built pinball machines using motors from tape players. Eventually, I learned that my main interests were in machines that moved.*

Arrick's early years with the company were spent building industrial robots that performed tasks such as moving test tubes and dispensing glue. Even though this paid the bills, he had a desire to offer a mobile robot, which spawned the design of the three-wheeled Trilobot (*www.robotics.com/trilobot/*). Trilobot found its way into university labs and high school classes, but it soon became obvious that a more affordable robot for hobbyists was in order–the ARobot was born.

Unlike many other kits designed for the hobby market, the ARobot uses a metal chassis and a real gear motor. It rapidly gained popularity among hobbyists and soon found itself featured on the front page of the Edmund Scientifics catalog (*www.scientificsonline.com*), one of the largest retailers of educational toys, hobby supplies, and scientific equipment. In 2003 Arrick released a book titled "Robot Building for Dummies" (*www.robotics.com/rbfd/book.html*), which taught the fundamentals of robot building using the ARobot as an example.

When asked what makes the ARobot so special and applicable for robotics applications designed with MSRS, Arrick responded with:

> *ARobot is inherently expandable and open by design. All low-level hardware functionality is accessible through an expansion connector, which is fully documented and supported by examples in the user guide.*
>
> *ARobot also utilizes a coprocessor dedicated to drive motor and steering control. This relieves the high-level controller from minutia, which often leads to an over-taxed processor and simplifies programming code that would otherwise be bogged down in the gory details of real-time hardware juggling.*
>
> *These qualities, along with a decent payload capacity, make ARobot perfect for use with a user-provided high-level controller running Microsoft Robotics Studio.*

You can purchase a complete ARobot kit from the Arrick Robotics Web site (*/www.robotics.com/arobot*) for $284. The ARobot will take you around three hours to assemble with basic hand tools. After you assemble it, the robot can be programmed using BASIC Stamp software provided by a company named Parallax (*www.parallax.com*). You can download a free version of the BASIC Stamp software from the following URL: *www.parallax.com/html_pages/downloads/software/software_basic_stamp.asp*. You need this to build the onboard remote communications interface for the ARobot. We will cover building this interface in the next section, "Build an Onboard Interface."

Unlike the robots covered in previous chapters, the ARobot does not have a two-wheel differential drive system. Instead, it uses a remote control servo motor to steer and a bidirectional, front-drive gear motor to move the robot. Because of this, we will not use the generic drive service included with MSRS when controlling the ARobot's ability to move. The steps for building the drive service will be covered in the section titled "Build a Drive Service."

The ARobot includes two front whisker sensors that can be used as bumpers for the robot. When triggered, they will indicate that an object collision is about to happen. In this case, we will be able to use the generic contact sensors service to implement a bumper service for the ARobot. This will be covered in the section titled "Build a Bumper Service."

To communicate with the ARobot using MSRS, we will need to build several layers or services that serve as an abstraction for the actual robot (see Figure 7-2). The first layer that is needed is the onboard interface, which consists of a BASIC program that executes on the ARobot's BS2 controller. This program will communicate directly with the ARobot's control service.

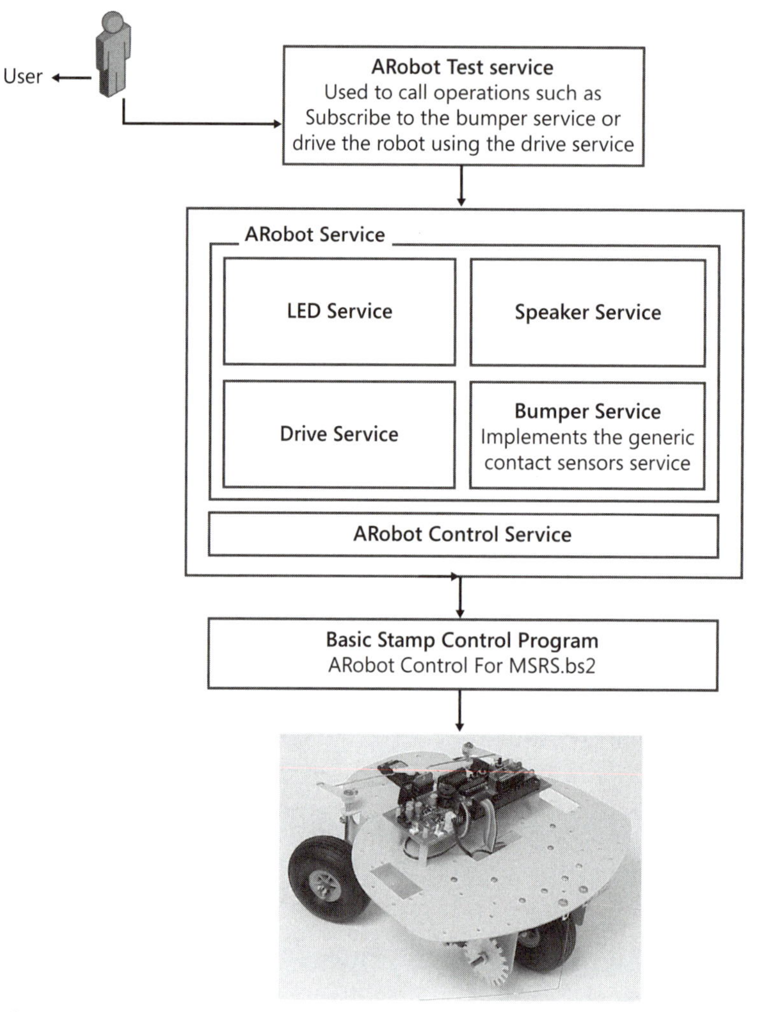

Figure 7-2 Diagram of the multilayered services used to communicate with the ARobot.

The core service is responsible for accepting operation requests from the ARobot services and mapping them to the appropriate commands that execute on the ARobot. For example, to make the robot move, you need to execute commands on the ARobot that operate the steering and drive motors. Operation requests are sent from the ARobot's drive service to the core service. These operations are then mapped to the appropriate commands and sent to the ARobot via a wired or wireless serial data connection. The result is movement by the robot.

Each sensor or actuator is represented through an individual service. Multiple high-level services are used to represent the interface that outside services use to communicate with the ARobot. If you were to add an additional component to the ARobot, such as an ultrasonic sensor, then an additional service representing this sensor would also need to be added. Only the high-level services should communicate directly with the core service. Services used to control the ARobot should access these services and not the core service.

Tip After assembling your ARobot, do not be surprised when you power it up and find that it does nothing. The BS2 comes to you empty. To get the robot to do anything, you will have to download a BASIC program to the coprocessor using the BASIC Stamp Editor. You can download the ARobot User Guide for detailed instructions on how to assemble and operate the ARobot through the following URL: *www.robotics.com/arobot/guide.html*.

Build an Onboard Interface

The onboard interface acts as a software driver. It is used to interpret commands sent to a remotely connected robot from an MSRS service. What constitutes the interface depends on the type of robot you are interfacing with. For example, the ARobot uses a BS2 as the coprocessor. The BS2 has only 32 bytes of memory, and only 26 of those are available to a downloaded program. Obviously, neither MSRS nor the .NET Framework is able to execute on this coprocessor. For this reason, we need software that can execute on the robot and pass information back and forth to a remote development machine capable of running MSRS.

The software we need is a BASIC program, which can be written and downloaded to the ARobot using a Basic Stamp Editor, such as the one available as a free download from the Parallax Web site (*www.parallax.com/html_pages/downloads/software/software_basic_stamp.asp*). The BASIC interface program is responsible for accepting commands from the MSRS services and returning information concerning the whisker sensors. The first thing the interface program does is make the speaker beep twice and turn on the green light-emitting diode (LED) light. This signals that the ARobot is ready to start accepting commands from MSRS. The first portion of this code should look like the following:

```
Start:
  ' Performs 2 short beeps when the robot starts up
  FREQOUT SPEAKER,150,2000
  PAUSE 10
  FREQOUT SPEAKER,150,2000
  LOW SPEAKER                        'turn off speaker.
  LOW GREENLED           'turn on the Green LED light
```

The next thing this and any interface program should do is begin a loop. Within the loop should be code that accepts input from the serial port and performs commands on the robot based on the data it receives. The loop should also contain code that returns data from sensors and actuators. For the interface program used by the ARobot, the code looks like the following:

```
Main:
'-----------------------------------------------------------------
' Main routine - this loop should run continuously
'-----------------------------------------------------------------
```

```
DO
  ' Get the next command by using the SERIN command
  ' to receive in the serial data sent in a packet
  ' The packet should contain no more than 4 bytes
  ' after the MSRS header
  SERIN Console, CMD_BAUD,300, NoCmd, [WAIT ("MSRS"), STR inBuffer \4]

  ' Process the incoming command by looking up what subroutine
  ' it is calling
  LOOKDOWN inBuffer(0), = [DRIVE_ROBOT, ON_RED_LED, OFF_RED_LED, ON_GREEN_LED,
OFF_GREEN_LED, SET_SPEAKER], routine
  ON routine GOSUB    DriveRobot, OnRedLED, OffRedLED, OnGreenLED, OffGreenLED, SetSpeaker

  PAUSE 100

 NoCmd: ' We want to always return whisker values

  outBuffer(0) = LEFT_WHISKER       'Left whisker
  outBuffer(1) = RIGHT_WHISKER      'Right Whisker
  outBuffer(2) = 0
  SEROUT Console, CMD_BAUD, ["ROB", STR outBuffer \4]

LOOP
```

Based on data it receives from MSRS, the interface program executes a subroutine. This subroutine contains specific functions such as driving the robot, setting the speaker, and turning on and off the LED lights. For example, the DriveRobot subroutine accepts parameters representing the distance, direction, and speed from the input buffer. Based on the values it receives, it sends commands to the BS2 coprocessor using SEROUT commands. The code for the DriveRobot subroutine is as follows:

```
DriveRobot:

  'First, get the input variables
  tDistance = inBuffer(1)
  tDirection = inBuffer(2)
  tSpeed = inBuffer(3)

  SELECT tDirection
    CASE "S"      'Stop the robot
      tmp1 = "0"
      tSteer(0) = "0"
      tSteer(1) = "0"
    CASE "B"      'Go backwards
      tmp1 = "0"
      tSteer(0) = "8"
      tSteer(1) = "0"
    CASE "F"      'Go forwards
      tmp1 = "1"
      tSteer(0) = "8"
      tSteer(1) = "0"
    CASE "R"      'Go right
      tmp1 = "1"
      tSteer(0) = "0"
```

```
    tSteer(1) = "1"
  CASE "L"      'Go left
    tmp1 = "1"
    tSteer(0) = "F"
    tSteer(1) = "F"
ENDSELECT

'Send commands to the CoProcessor and get a response
SEROUT COPROC, NET_BAUD, ["!1R1"]       'Send RC command to coprocessor.
SEROUT COPROC, NET_BAUD, [tSteer(0)]    'Steer command 1
SEROUT COPROC, NET_BAUD, [tSteer(1)]    'Steer command 2
SERIN COPROC, NET_BAUD,  [tResponse]    'Get a response
PAUSE 300                               'Pause 300 ms

SEROUT COPROC, NET_BAUD, ["!1M1"]       'Starting part of command to co.
SEROUT COPROC, NET_BAUD, [tmp1]         'Set the direction 1=forward 0=backward
SEROUT COPROC, NET_BAUD, [HEX1 tSpeed]  'Set the speed
IF tDistance = 0 AND tDirection = "S" THEN
  SEROUT COPROC, NET_BAUD, ["0001"]     'Turn the motor off
ELSEIF tDistance = 0 AND tDirection <> "S" THEN
  SEROUT COPROC, NET_BAUD, ["FFFF"]     'Set Distance for forever
ELSE
  SEROUT COPROC, NET_BAUD, [HEX4 tDistance] 'Set Distance set number of inches
ENDIF

SERIN COPROC, NET_BAUD, [tResponse]     'Get a response
PAUSE 100                               'Pause 100 ms

RETURN
```

Note You may be a bit confused when you notice that the code is sending serial commands to another processor. This is because the ARobot has two controllers. One is the main controller board, and this is where you send the serial commands from the MSRS service. The main controller board is used to control the robot's speaker and LED lights. Additionally, there is a BS2 controller, which is used to power the RC steering and gear motors.

The remaining subroutines are used to turn on and off the LED lights and set the speaker using duration and frequency parameters passed into the input buffer. The code for these subroutines is as follows:

```
OnRedLED:

  LOW REDLED           'turn on the Red LED light
  RETURN

OffRedLED:

  HIGH REDLED          'turn off the Red LED light
  RETURN

OnGreenLED:
```

```
  LOW GREENLED         'turn on the Green LED light
  RETURN

OffGreenLED:

  HIGH GREENLED        'turn off the Green LED light
  RETURN

SetSpeaker:

  tDuration = inBuffer(1)    'Get time in ms that speaker should sound out
  tFrequency = inBuffer(2)   'get the frequency
  FREQOUT SPEAKER, tDuration * 50, tFrequency * 50   'Set the tone
  LOW SPEAKER          'turn off the speaker

  RETURN
```

The code for the onboard interface program is available on this book's companion Web site. The file named ARobotControlForMSRS.bs2 should be loaded into the Basic Stamp Editor and executed by clicking Run from the program menu bar (see Figure 7-3). If the download was successful, you should hear two short beeps, and the green LED light will turn on.

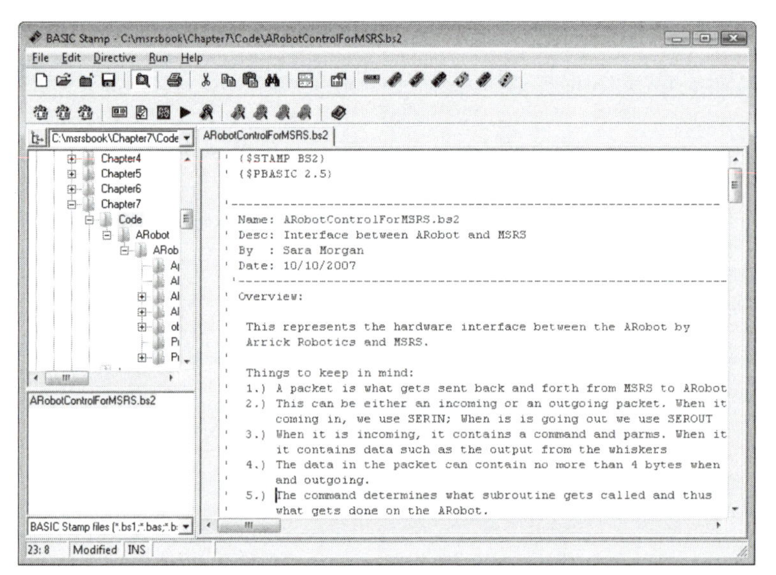

Figure 7-3 The Basic Stamp Editor is used to download and execute the onboard interface program.

Build a Core Service

The next step in building a hardware interface is to design a core service. This service is also commonly referred to as the brick service. This is a Decentralized Software Services (DSS) service that serves to represent the hardware–or, in this case, the ARobot. The core service is responsible for communicating with the ARobot through a serial port. You can connect the ARobot to the development machine through either a wired or a wireless connection.

> **Note** The ARobot does not come packaged with wireless capabilities. To communicate with the ARobot wirelessly, you will have to install a wireless receiver on the ARobot.

To create the core service, create a new DSS service using the Visual Studio template provided with MSRS. You can name the service ARobot. By default, this creates two source files: ARobot.cs and ARobotTypes.cs. The ARobotTypes.cs code file is where you add code concerning the robot's state and list the DSS operations that can be performed.

Add Code to the Contract Class

The ARobot controller board comes with red and green LED lights, whiskers, and a speaker. Each of these components is represented through state variables. There is also a state variable used to hold the COM port number. The COM port number is assigned when you connect the ARobot to your computer using a serial cable.

Additionally, we need state variables to represent the configuration parameters used to drive the robot. These are variables representing direction, speed, and distance. For these variables to be returned with the robot's state, they must be declared in the *ARobotTypes* contract class and include the *DataMember* attribute above each variable declaration. For example, you can use the following code to represent the state for the ARobot:

```
[DataContract()]
public class ARobotState
{

    private bool _connected;
    private int _comPort;
    private DriveConfig _driveConfig;
    private Whiskers _whiskers;
    private Speaker _speaker;
    private LEDs _leds;

    [DataMember]
    [Description("Indicates whether the ARobot is connected.")]
    public bool Connected
    {
        get { return this._connected; }
        set { this._connected = value; }
    }

    [DataMember]
    [Description("Indicates COM Port that the ARobot is connected to.")]
    public int ComPort
    {
       get { return this._comPort; }
       set { this._comPort = value; }
    }

    [Description("Drive Configuration")]
    [DataMember]
```

```
    public DriveConfig DriveConfig
    {
        get { return this._driveConfig; }
        set { this._driveConfig = value; }
    }

    [Description("Whisker State")]
    [DataMember]
    public Whiskers Whiskers
    {
        get { return this._whiskers; }
        set { this._whiskers = value; }
    }

    [Description("Speaker Configuration")]
    [DataMember]
    public Speaker Speaker
    {
        get { return this._speaker; }
        set { this._speaker = value; }
    }

    [Description("LED State")]
    [DataMember]
    public LEDs LEDs
    {
        get { return this._leds; }
        set { this._leds = value; }
    }
}
```

The user-defined types referenced in the state variable declarations (i.e. DriveConfig, Whiskers, Speaker, and LEDs) are defined beneath the state declarations. These class definitions must also use the *DataMember* attribute, along with the *DataContract* attribute. For example, the class definition used to define the DriveConfig type is shown as follows:

```
[Description("The ARobots Drive configuration parameters.")]
[DataContract]
public class DriveConfig
{
    private int _motorSpeed;
    private int _distance;
    private string _direction;

    [DataMember]
    [Description("Indicates the motor speed used by the ARobot.")]
    public int MotorSpeed
    {
        get { return this._motorSpeed; }
        set { this._motorSpeed = value; }
    }

   [Description("Indicates the direction that the ARobot moves towards.")]
   [DataMember]
```

```
    public string Direction
    {
        get { return this._direction; }
        set { this._direction = value; }
    }
    [Description("Indicates the distance in half-inches ARobot drives")]
    [DataMember]
    public int Distance
    {
        get { return this._distance; }
        set { this._distance = value; }
    }
    public DriveConfig()
    {
    }

    public DriveConfig(int motorSpeed, string direction, int distance)
    {
        MotorSpeed = motorSpeed;
        Direction = direction;
        Distance = distance;
    }
}
```

You can see that the ARobot uses three input parameters to drive the robot: MotorSpeed, Direction, and Distance. These variables are set with values that are passed on to the ARobot when a request is made to drive the robot. The following value ranges can be used for these variables:

- **MotorSpeed** Use a value from 0 to 15, in which 0 indicates no movement and 15 is the maximum speed.
- **Direction** Use a string-based value that represents one of the following: F=Forwards, B=Backwards, R=Right, L=Left, S=Stop.
- **Distance** Use an integer-based value that represents the distance the ARobot will travel. You can use an approximate value of 16 to represent 12 inches or 1 foot. If you wanted the robot to move 2 feet, then you would use a distance of 32.

Modify the Portset

Based on the interface program created in an earlier section, the ARobot can accept commands to drive the robot, turn on and off the lights, activate the speaker, and receive sensor information from the whiskers. These functions need to be represented as DSS operations. You define them in the *ARobotTypes* class and add them to the PortSet using the following code:

```
[ServicePort()]
public class ARobotOperations : PortSet<DsspDefaultLookup,
                           DsspDefaultDrop,
                           Get,
```

```
                        Replace,
                        Subscribe,
                        UpdateWhiskers,
                        PlaySpeaker,
                        SetLEDs,
                        DriveRobot>
{
}
```

As you learned in earlier chapters, the *DsspDefaultLoopkup*, *DsspDefaultDrop*, and *Get* operations are standard DSS operations added to the code when the template is created. The *Replace* operation is added to allow for a change to the state. The *Subscribe* operation allows us to receive notifications whenever the whiskers state has changed. The remaining operations correspond to the functions that can be performed on the ARobot.

Add Code to the Implementation Class

The implementation class, which in this case is named *ARobot*, is where you add code used to send commands to the ARobot onboard interface. To do this, you first need to add code that declares the subscription manager service as a partner. This is needed when we implement the code associated with the subscribe operation. The code to declare the subscription service partner is as follows:

```
   [Partner(dssp.Partners.SubscriptionManagerString,
      Contract = submgr.Contract.Identifier,
      CreationPolicy = PartnerCreationPolicy.CreateAlways)]
submgr.SubscriptionManagerPort _subMgrPort = new
   submgr.SubscriptionManagerPort();
```

You also need to add a declaration for the Initial State partner. This allows us to retrieve state values from an XML-based file, which in this case is a file named ARobot.config.xml. The code for this declaration should appear as follows:

```
[InitialStatePartner(Optional = true, ServiceUri =
         "samples/config/arobot.config.xml")]
private ARobotState _state = new ARobotState();
```

> **Note** By default, MSRS searches for the Initial State partner in the store directory beneath the MSRS installation folder. If you want to move the config file to a different directory (such as the samples/config directory), you need to include a portion of the path in the Initial State partner declaration.

The ARobot.config.xml file contains values for each state variable included in the contract class. The ARobot.config.xml file should look like the following:

```
<?xml version="1.0" encoding="utf-8"?>
<ARobotState xmlns:s=http://www.w3.org/2003/05/soap-envelope
  xmlns:wsa="http://schemas.xmlsoap.org/ws/2004/08/addressing"
```

```
  xmlns:d="http://schemas.microsoft.com/xw/2004/10/dssp.html"
  xmlns="http://custsolutions.net/2007/10/arobot.html">
<Connected>false</Connected>
<ComPort>4</ComPort>
<DriveConfig>
<Direction>F</Direction>
<MotorSpeed>3</MotorSpeed>
<Distance>0</Distance>
</DriveConfig>
<Whiskers>
<WhiskerLeft>false</WhiskerLeft>
<WhiskerRight>false</WhiskerRight>
</Whiskers>
<Speaker>
<frequency>50</frequency>
<Duration>250</Duration>
</Speaker>
<LEDs>
<RedLED>false</RedLED>
<GreenLED>false</GreenLED>
</LEDs>
</ARobotState>
```

Modify the *Start* Method

The next thing to do is add code to the *Start* method. The *Start* method is created through the template when the service is created. You need to add code to the *Start* method so that it looks like the following:

```
protected override void Start()
{
    base.Start();

    ConnectToARobot();

    //add custom handlers to interleave
    Activate(
     Arbiter.Interleave(
      new TeardownReceiverGroup(),
      new ExclusiveReceiverGroup(
        Arbiter.ReceiveWithIterator<SetLEDs>(true, _mainPort,
                SetLEDsHandler),
        Arbiter.ReceiveWithIterator<UpdateWhiskers>(true, _mainPort,
                UpdateWhiskersHandler),
        Arbiter.ReceiveWithIterator<DriveRobot>(true, _mainPort,
                SetDriveConfigHandler),
        Arbiter.ReceiveWithIterator<PlaySpeaker>(true, _mainPort,
                PlaySpeakerHandler)
         ),
      new ConcurrentReceiverGroup()
    ));
 }
```

Connect to the ARobot

The *ConnectARobot* method sets the value for the *Connected* state variable by calling the *Connect* function. The *Connected* state variable indicates whether MSRS has successfully connected to the ARobot. The code for this method looks like the following:

```
private void ConnectToARobot()
{
    //Connect to the ARobot and establish connection which will
    //register a receive handler to get the whisker data coming
    //back from the robot
    string errorMessage;
   _state.Connected = Connect(_state.ComPort, out errorMessage);
    if (!_state.Connected && !string.IsNullOrEmpty(errorMessage))
    {
        LogError(LogGroups.Activation, errorMessage);
        return;
    }
}
```

Before you can run the ARobot services, you need to connect your development machine to the ARobot using either a serial cable or a wireless connection. You get the number for the port assignment at the time you connect your machine to the robot. After you get this port assignment number, change the value for the state variable named *ComPort*. This value is retrieved from the ARobot.config.xml file. Open the file with a text editor and change the associated value.

The *Connect* function first checks to see that we are using a valid COM port number. This number is determined at the point a serial connection to the ARobot is established on your development machine; therefore, it varies with each machine. The code for the *Connect* function is shown as follows:

```
public bool Connect(int serialPort, out string errorMessage)
{
      //Make sure we have a valid COM port number
      if (serialPort <= 0)
      {
          errorMessage = "The ARobot serial port is not configured!";
          return false;
      }

      //Check to see if we are already connected
      if (connected)
      {
          Close();
      }

      //Open a connection
      try
      {
          string ComPort = "COM" + serialPort.ToString();
          _serialPort = new SerialPort(ComPort, _baudRate);
```

```
            _serialPort.Parity = Parity.None;
            _serialPort.DataBits = 8;
            _serialPort.StopBits = StopBits.One;
            _serialPort.ReadTimeout = 2000; //in ms
            _serialPort.WriteTimeout = 2000; //in ms
            _serialPort.ReceivedBytesThreshold = 1;
            _serialPort.DataReceived += new
                 SerialDataReceivedEventHandler(_serialPort_DataReceived);
            _serialPort.Open();
            _serialPort.DiscardInBuffer();
            errorMessage = string.Empty;
            connected = true;
            return true;
        }
        catch (Exception ex)
        {
            errorMessage = string.Format("Error connecting ARobot to
                   COM{0}:1}", serialPort, ex.Message);
            return false;
        }
    }
```

Create Message Handlers

The ARobot service uses four message handlers to handle the associated DSS operations listed in the PortSet. For example, there is a message handler named *UpdateWhiskersHandler* that is called every time the whisker state variables are updated. The code for this handler is shown as follows:

```
public virtual IEnumerator<ITask> UpdateWhiskersHandler(UpdateWhiskers update)
{
     //Value the state variables
     _state.Whiskers.WhiskerLeft = update.Body.WhiskerLeft;
     _state.Whiskers.WhiskerRight = update.Body.WhiskerRight;

     // Send Notifications to subscribers
     SendNotification<UpdateWhiskers>(_subMgrPort, new
            UpdateWhiskers(_state.Whiskers));

     //Post the update operation
     update.ResponsePort.Post(DefaultUpdateResponseType.Instance);
     yield break;
}
```

Before updating the value of the whisker state variables, a notification is sent to the subscription port, which indicates that the whisker state has changed. All subscribers to the service are then notified of the change. The remaining handlers are used to process updates associated with the robot's speaker, LEDs, and drive configuration. Each of these handlers invokes a method that sends commands to the ARobot's serial port. For example, the *SetDriveConfigHandler* is called whenever the *DriveRobot* operation is called. This handler calls another method named *DriveRobot* and passes to it variables representing the distance, direction, and motor speed. The *SetDriveConfigHandler* method is shown as follows:

```
public virtual IEnumerator<ITask> SetDriveConfigHandler(DriveRobot update)
{
     //Make sure we are connected
     if (!_state.Connected)
     {
          LogError("trying to drive the robot, but not connected");
          update.ResponsePort.Post(new Fault());
          yield break;
     }

     //Value the state variables
     _state.DriveConfig.Distance = update.Body.Distance;
     _state.DriveConfig.Direction = update.Body.Direction;
     _state.DriveConfig.MotorSpeed = update.Body.MotorSpeed;

     //Send command to the ARobot to set the configuration and drive robot
     DriveRobot(update.Body.Distance, update.Body.Direction,
         update.Body.MotorSpeed);

     //Post the update operation
     update.ResponsePort.Post(DefaultUpdateResponseType.Instance);
     yield break;
}
```

The *DriveRobot* method is used to send the drive configuration information to the ARobot through a byte array. The following code shows what the *DriveRobot* method looks like:

```
public void DriveRobot(int inches, string direction, int speed)
{

     byte[] packet = new byte[_packetSize];
     packet[0] = _driveRobot;
     //Approximate distance in half inches
     packet[1] = Convert.ToByte(inches);
     //Direction should be a S=Stop,B=Backwards,F=Forwards,R=Right,L=Left
     packet[2] = (byte)Convert.ToChar(direction);
  //Motor Speed is set with value ranging from 0 to 15
     packet[3] = Convert.ToByte(speed);
     SendData(ref packet);
}
```

The byte array is necessary because this is what the BS2 accepts through the serial port. Depending on what type of robot you are designing an interface for, the method of interfacing with that robot may vary. For the ARobot interface, the first byte in the array contains a unique number that represents a command for the robot. In the case of the DriveRobot command, this number is 20. Why 20? The number was selected when the interface was designed, and it is used by the onboard interface (introduced in the last section) to determine which subroutine is executed. Constants located at the top of the ARobot class file store the values for these command assignments. If for some reason the numbers assigned to these commands are changed in the onboard interface program, you must also change the associated constant value in the *ARobot* class.

The last step is to add the remaining handlers, which represent updates to the speaker and LEDs. After this is done, you can compile the ARobot project by clicking Build and then clicking Build Solution from the menu bar. In this chapter, we only cover how to create the drive and bumper services. The LED and speaker services, which are included with the services on the companion Web site, function much the same as the drive service.

Tip If you encounter strange errors when trying to build the proxy project for your service, try compiling manually using the msbuild command. This is done by going to the command prompt for MSRS and typing something similar to the following: **msbuild "samples\myservice\myservice.csproj."**

Build a Drive Service

Each of the services representing the robot's sensors and actuators should be contained within a new Visual Studio project. This project can be added to the ARobot solution, but the class files that represent each service should reside within a project separate from the class files for the control service. To do this, you can right-click the ARobot solution in Solution Explorer, click Add, and then click New Project. The project should be named ARobot-Services, and you should create it using the Simple Dss Service template in Visual Studio.

Set References

After you have added the project to the solution, you need to set a reference to the ARobot control service. When the control service was compiled in the last section, four assembly files were created and added to the bin folder for your local MSRS installation. You need to add a reference to the proxy assembly for the ARobot service by right-clicking the References folder in Solution Explorer and choosing Add Reference. From the .NET tab, select the assembly named *ARobot.Y2007.M10.Proxy* and click OK (see Figure 7-4).

Note The *ARobot.Y2007.M10.Proxy* assembly might be named something different if you created the ARobot project from scratch and did not use the code provided on the companion Web site. If this is the case, you need to reference the name of the proxy file that was created when you compiled the service in the last section.

Add Reference

.NET | COM | Projects | Browse | Recent

Component Name	Version	Runtime	Pat
Arcos.Y2006.M06.proxy	0.0.0.0	v2.0.50727	c:\
Arcos.Y2006.M06.transform	0.0.0.0	v2.0.50727	c:\
ARobot.Y2007.M10	0.0.0.0	v2.0.50727	c:\
ARobot.Y2007.M10.Proxy	0.0.0.0	v2.0.50727	c:\
ARobot.Y2007.M10.transform	0.0.0.0	v2.0.50727	c:\
ARobotServices.Y2007.M10	0.0.0.0	v2.0.50727	c:\
ARobotServices.Y2007.M10.Proxy	0.0.0.0	v2.0.50727	c:\
ARobotServices.Y2007.M10.transform	0.0.0.0	v2.0.50727	c:\
ARobotTest.Y2007.M10	0.0.0.0	v2.0.50727	c:\
ARobotTest.Y2007.M10.Proxy	0.0.0.0	v2.0.50727	c:\

OK | Cancel

Figure 7-4 The Add References dialog box is used to add a reference to the ARobot control service.

> **Tip** When adding references to services, set the Copy Local property to False. This prevents the DLL for the proxy service to be copied to the local output directory. You want assemblies to remain in the bin folder for the local MSRS installation. This can prevent problems down the line with strong name conflicts.

Add a Contract Class

When you create a DSS service using the Visual Studio template, it creates contract and implementation classes for you by default. For the ARobotServices project, we need to rename these files as follows:

- ARobotServices.cs becomes ARobotDrive.cs
- ARobotServicesTypes.cs becomes ARobotDriveTypes.cs

The contract class for the drive service is the ARobotDriveTypes class file. The namespace used in this class should be the following:

```
Namespace Microsoft.Robotics.Services.ARobot.Drive
```

Additionally, the contract class should contain a string constant for the identifier. You should set it with a unique URI, such as in the following contract class definition:

```
public sealed class Contract
{
   public const string Identifier =
          "http://custsolutions.net/2007/10/arobotdrive.html";
}
```

> **Tip** Notice the use of custsolutions.net in the contract identifier. By default, an identifier uses schemas.tempuri.org as part of the identifier path. To keep this identifier unique, it is recommended that you change this portion of the identifier.

It is not necessary for your service to have a state variable, but you may wish to include something that indicates whether the service is being used. For the drive service, we use a state variable named *DrivingRobot*. This variable can be added to the state using the following code:

```
[DataContract()]
public class ARobotDriveState
{
   private bool _drivingRobot;

   public bool DrivingRobot
   {
       get { return this._drivingRobot; }
       set { this._drivingRobot = value; }
   }
}
```

The next thing to add is the PortSet, which indicates what DSS operations can be performed on the port. For the drive service, we need to add only an operation named *DriveRobot*. The code defining the PortSet should look like the following:

```
public class DriveRobotOperations : PortSet<
        DsspDefaultLookup,
        DsspDefaultDrop,
        Get,
        DriveRobot>
{
}
```

You also need to add a class definition, which represents a DriveRobot request. The request object holds the parameter values needed to operate the robot. The class definition must include the *DataContract* and *DataMemberConstructor* attributes. Each member variable must include the *DataMember* attribute. The code for the DriveRobot request should look like the following:

```
[DataContract]
[DataMemberConstructor]
public class DriveRobotRequest
{
    private int _motorSpeed;
    private int _distance;
    private string _direction;
```

```
    [DataMember]
    public int MotorSpeed
    {
        get { return this._motorSpeed; }
        set { this._motorSpeed = value; }
    }

    [DataMember]
    public string Direction
    {
        get { return this._direction; }
        set { this._direction = value; }
    }

    [DataMember]
    public int Distance
    {
        get { return this._distance; }
        set { this._distance = value; }
    }
}
```

Tip If you forget to add the *DataMember* attribute to the member variables of the *Drive-RobotRequest* class, then those variables are not available to a service trying to partner with the drive service.

Add an Implementation Class

You need to add namespace references for the ARobot control service to the implementation class. This is done by adding the following code to the top of the ARobotDrive class file:

```
using arobot = Microsoft.Robotics.Services.ARobot.Proxy;
```

Because the *ARobot* control class is responsible for sending commands to the robot, you need to add it as a partner service. You also need to declare a port in which ARobot operations can be posted to. For the drive service, this port is named _aRobotPort. The following code shows what this declaration looks like:

```
[Partner("ARobot", Contract = arobot.Contract.Identifier,
       CreationPolicy = PartnerCreationPolicy.UseExistingOrCreate, \

       Optional = false)]
private arobot.ARobotOperations _aRobotPort = new
         arobot.ARobotOperations();
```

The *Start* method needs to initialize the only state variable used by the drive service. The code for the *Start* method is shown as follows:

```
protected override void Start()
{
    base.Start();

    //Initialize the state variable
    if (_state == null)
    {
        _state = new ARobotDriveState();
        _state.DrivingRobot = false;
    }
}
```

The *DriveRobot* message handler is responsible for posting the DriveRobot request to the _aRobotPort and, thereby, invoking the DriveRobot in the *ARobot* class. The following code for this message handler is shown as follows:

```
[ServiceHandler(ServiceHandlerBehavior.Exclusive)]
public virtual IEnumerator<ITask> DriveRobot(DriveRobot request)
{

     arobot.DriveConfig drive = new arobot.DriveConfig();
     drive.Direction = request.Body.Direction;
     drive.Distance = request.Body.Distance;
     drive.MotorSpeed = request.Body.MotorSpeed;

     //Update the State Variable which indicates whether
     //the robot is being driven
     if (drive.Direction == "S")
     {
         _state.DrivingRobot = false;
     }
     else
     {
         _state.DrivingRobot = true;
     }

    yield return Arbiter.Choice(_aRobotPort.DriveRobot(drive),
          delegate(DefaultUpdateResponseType response)
          {
                request.ResponsePort.Post(
                     DefaultUpdateResponseType.Instance);
          },
          delegate(Fault fault)
          {
              request.ResponsePort.Post(new Fault());
          }
    );

    yield break;
}
```

Before you can compile the ARobotServices project, you need to add classes used to represent the remaining services: bumper, LED, and speaker. In the next section, we cover the steps for building the bumper service.

Build a Bumper Service

The bumper service is different from the drive, LED, and speaker services because it implements the generic contact sensors contract. The benefit for using the generic contact sensors service is that we need only one class file to represent the bumper service. The amount of code used to represent the bumper service is reduced, and we can use the state variables and DSS operations that are part of the generic contact sensors service. To add the class file, right-click the ARobotServices project, click Add, and then click Class. You can name the new class file ARobotBumper.

Add a Reference

The generic contact sensors service is part of the *Robotics.Common* assembly. This means you need to add a reference to the *Robotics.Common.Proxy* assembly using the Add References dialog box. You also need to add namespace references for the ARobot control service and generic contact sensors service. Additionally, you need to add a namespace reference for the subscription manager service because receiving notifications of whisker changes involves creating a subscription. The code for these namespace references is shown as follows:

```
using arobot = Microsoft.Robotics.Services.ARobot.Proxy;
using bumper = Microsoft.Robotics.Services.ContactSensor.Proxy;
using submgr = Microsoft.Dss.Services.SubscriptionManager;
```

The namespace for this service is *Microsoft.Robotics.Services.ARobot.Bumper*. Because we are using the generic contact sensors contract and this comes from another assembly, we need to use the *AlternateContract* attribute. The attributes listed at the top of the *ARobotBumper* class should look like the following:

```
[Contract(Contract.Identifier)]
[AlternateContract(bumper.Contract.Identifier)]
[DisplayName("ARobot Generic Contact Sensor")]
[Description("Provides access to the ARobot whisker used as a
            bumper.\n(Uses Generic Contact Sensors contract.)")]
```

We still need a contract identifier for the ARobotBumper service. This is defined with the following class declaration and added to the top of the ARobotBumper class file:

```
public sealed class Contract
{
   public const string Identifier =
        "http://custsolutions.net/2007/10/arobotbumper.html";
}
```

Add Partnerships

The bumper service declares an initial state partner using the *InitialStatePartner* attribute. This indicates that the generic contact sensors service contains the state used to initialize the bumper service. The code for this declaration is as follows:

```
[InitialStatePartner(Optional = true)]
private bumper.ContactSensorArrayState _state = new
      bumper.ContactSensorArrayState();
```

Additionally, we need to declare a main port for the bumper service, as well as ports for the ARobot control service and subscription manager service. The code for these declarations is as follows:

```
[ServicePort("/ARobotBumper", AllowMultipleInstances = true)]
private bumper.ContactSensorArrayOperations _mainPort = new
      bumper.ContactSensorArrayOperations();

[Partner("ARobot", Contract = arobot.Contract.Identifier,
CreationPolicy = PartnerCreationPolicy.UseExistingOrCreate,
Optional = false)]
private arobot.ARobotOperations _aRobotPort = new
arobot.ARobotOperations();

[Partner(dssp.Partners.SubscriptionManagerString, Contract = submgr.Contract.Identifier,
CreationPolicy = PartnerCreationPolicy.CreateAlways,
Optional = false)]
private submgr.SubscriptionManagerPort _subMgrPort = new
      submgr.SubscriptionManagerPort();
```

In the code for the *Start* method, we will initialize the state for the contact sensor array and then use the subscribe operation to receive whisker notifications. The code for the *Start* method will look like the following:

```
protected override void Start()
{
    InitializeState();

    // Subscribe to the ARobot for whisker notifications
    SubscribeToARobot();

   base.Start();
}
```

The *SubscribeToARobot* method is responsible for creating a notification port and then sending a Subscribe request to the ARobot notification port. It uses the *Arbiter.Receive* method to indicate that the service should wait for a response, and, if the response is an error, then it logs that error. The code for the *SubscribeToARobot* method is as follows:

```
private IEnumerator<ITask> SubscribeToARobot()
{
     // Create a notification port
```

```
    arobot.ARobotOperations notificationPort = new
          arobot.ARobotOperations();

    Fault f = null;
    // Subscribe to the ARobot and wait for a response
    yield return Arbiter.Choice(_aRobotPort.Subscribe(notificationPort),
         delegate(SubscribeResponseType Rsp)
         {    },
         delegate(Fault fault)
         {
            LogError("Subscription failed", fault);
         }
    );

    base.MainPortInterleave.CombineWith(new Interleave(
       new ExclusiveReceiverGroup(
         Arbiter.Receive<arobot.UpdateWhiskers>(
           true, notificationPort, SensorsChangedHandler)
         ), new ConcurrentReceiverGroup()
    ));
}
```

The *SubscribeToARobot* method registers a message handler when it subscribes to the ARobot. This message handler, named *SensorChangedHandler*, is responsible for determining which sensor has been activated and then updating the generic contact sensor property named *Pressed*. This property indicates whether one of the contact sensors has been pressed. It also sends a notification to the subscription manager port, which indicates that all subscribers should be notified of the update. The code for the *SensorsChangedHandler* is as follows:

```
private void SensorsChangedHandler(arobot.UpdateWhiskers notification)
{
     bool changed = false;
     bumper.ContactSensor bumper = null;

     for (int i = 0; i < _state.Sensors.Count; i++)
     {
        bumper = _state.Sensors[i];

        if (bumper.HardwareIdentifier == 1 &&
              bumper.Pressed != notification.Body.WhiskerLeft)
        {
            bumper.Pressed = (notification.Body.WhiskerLeft);
            changed = true;
            break;
        }

        if (bumper.HardwareIdentifier == 2 &&
              bumper.Pressed != notification.Body.WhiskerRight)
        {
            bumper.Pressed = (notification.Body.WhiskerRight);
            changed = true;
            break;
```

```
            }

        }

        if (changed)
        {
            this.SendNotification<bumper.Update>(_subMgrPort, new
                  bumper.Update(bumper));
        }

    }
```

After all the changes are made to the ARobotBumper service and you have added class files representing the LED and speaker services, you can compile the ARobotServices project and create the assembly files in the bin folder for the MSRS installation. The next step is to create a test service that can use these new services to control the ARobot.

Build a Test Service

Now that you have built and successfully compiled the services for the ARobot, you can create a test service that consumes these services. The test service project can be added to the existing ARobot solution or created in a new solution. The code included on the book's companion Web site uses both a test service embedded within the ARobot solution and a VPL project. The VPL project is named DriveARobotByWire, and it is located in a separate folder named DriveARobotByWire.

Build the ARobotTest Project

The test service project is named ARobotTest, and it resides within the ARobot folder. The test service is simple and is used to test both the bumper and drive services. To access these services, we need to add a reference to the *ARobotServices.Y2007.M10.Proxy* assembly. We can then add namespace references to the drive service using the following code:

```
using drive = Microsoft.Robotics.Services.ARobot.Drive.Proxy;
```

We need to add a partnership for the generic contact sensors service in order to access the contact sensor array operations. The generic contact sensors service is part of the *Robotics.Common* assembly, so we need to set a reference to the *Robotics.Common.Proxy* assembly. We also need to add the following namespace declaration:

```
using contactsensor = Microsoft.Robotics.Services.ContactSensor.Proxy;
```

Notifications received from the bumper service trigger a Windows alert dialog box. To access this functionality, we need to set a reference to the *Utility.Y2006.M08.Proxy* assembly and include the following namespace declaration:

```
using dialog = Microsoft.Robotics.Services.Sample.Dialog.Proxy;
```

Finally, we need to include a namespace declaration that references the subscription manager service that is part of DSS. This is done with the following code:

```
using submgr = Microsoft.Dss.Services.SubscriptionManager;
```

Add Partnerships

The test service needs to add partnerships that represent the subscription manager service, Windows dialog box operations, generic contact sensors, and ARobot drive services. You can add these with the following code:

```
[Partner("SubMgr", Contract = submgr.Contract.Identifier,
      CreationPolicy = PartnerCreationPolicy.CreateAlways)]
private submgr.SubscriptionManagerPort _subMgr = new
      submgr.SubscriptionManagerPort();

[Partner("SimpleDialog", Contract = dialog.Contract.Identifier,
      CreationPolicy = PartnerCreationPolicy.UsePartnerListEntry)]
dialog.DialogOperations _simpleDialogPort = new dialog.DialogOperations();

[Partner("ARobotGenericContactSensor",
   Contract = contactsensor.Contract.Identifier,
      CreationPolicy = PartnerCreationPolicy.UsePartnerListEntry)]
contactsensor.ContactSensorArrayOperations _aRobotGenericContactSensorPort
        = new contactsensor.ContactSensorArrayOperations();
contactsensor.ContactSensorArrayOperations
       _aRobotGenericContactSensorNotify = new
       contactsensor.ContactSensorArrayOperations();

[Partner("ARobotDrive", Contract = drive.Contract.Identifier,
      CreationPolicy = PartnerCreationPolicy.UseExistingOrCreate,
      Optional = false)]
private drive.DriveRobotOperations _drivePort = new
      drive.DriveRobotOperations();
```

Modify the *Start* Method

The *Start* method for the test service subscribes to the generic contact sensors service. It also registers a message handler that is called whenever the ARobot's whiskers are triggered. Finally, use the *SpawnIterator* method to invoke a method named *MoveForwardAndStop*. The code for the *Start* method looks like the following:

```
protected override void Start()
{
base.Start();

   //Subscribe to the generic contact sensors service for the ARobot
   //This will get us access to the whiskers
_aRobotGenericContactSensorPort.Subscribe(_aRobotGenericContactSensorNotify);
```

```
base.MainPortInterleave.CombineWith(
    new Interleave(
       new ExclusiveReceiverGroup(
       ),
       new ConcurrentReceiverGroup(
          Arbiter.ReceiveWithIterator<contactsensor.Update>(true,
               _aRobotGenericContactSensorNotify,
               ARobotGenericContactSensorUpdateHandler)
       )
    )
 );

 SpawnIterator(MoveForwardAndStop);

}
```

Add Message Handlers

The contact sensors message handler is responsible for displaying the word "Ouch" in a Windows alert dialog box every time one of the ARobot's whiskers are triggered. The code for this message handler is as follows:

```
IEnumerator<ITask> ARobotGenericContactSensorUpdateHandler(contactsensor.Update message)
{
   //Display the word "Ouch" in a dialog box every time whisker is touched
   dialog.AlertRequest ar = new dialog.AlertRequest();
   ar.Message = "Ouch";
   _simpleDialogPort.Alert(ar);
   yield break;
}
```

Additionally, you need to add code for the *MoveForwardAndStop* method. This method is called when the service is started, and it is responsible for moving the robot forward for one second and then stopping. This is done by sending a DriveRobot request using a distance of 0, a speed of 6, and a direction of "F" to move the robot forward. To stop the robot, a direction of "S" is used. The code for the *MoveForwardAndStop* method is as follows:

```
private IEnumerator<ITask> MoveForwardAndStop()
{
    //Send a command to drive forward for no specific distance
    _drivePort.DriveRobot(new drive.DriveRobotRequest(6, "F", 0));

    // Wait for 1 second
    yield return Arbiter.Receive(false, TimeoutPort(1000),
          delegate(DateTime t) { });

    //Now Stop
    _drivePort.DriveRobot(new drive.DriveRobotRequest(3, "S", 0));

}
```

After you have added the code, you can compile and run the test service. Make sure the ARobot is connected to your development machine with a serial connection and the ARobotControlForMSRS.bs2 program has been loaded onto the ARobot. The result of running the test service is that the ARobot moves forward for one second and then stops. If the whiskers are triggered, a Windows alert dialog box will appear with the word "Ouch."

> **Tip** If you have issues with starting a service or sending messages, try the following steps to resolve the problem:

1. Delete the contractdirectory.cache.xml file, which is located in the store folder for your local MSRS installation.
2. Make sure only one version of your assembly, proxy, and transform files is in the bin folder for the local MSRS installation.
3. Make sure you do not have multiple DLLs claiming to implement the same contract. This can happen if you change the assembly name for your service after you created the initial service.
4. Make sure the DLLs for your service appear in the bin folder for the local MSRS installation.

Build the DriveARobotByWire Project

The VPL project is much simpler to write because it involves dragging objects onto a design surface and then setting properties related to those objects. The first step is to open VPL and drag an instance of the DirectionDialog service onto the design surface. You will use the direction dialog, which is a special service available through MSRS, to prompt the user for a direction. You then use a Calculate activity to interpret the name value entered into the direction dialog.

The Calculate activity captures only the name that is entered. A Switch activity, which functions like a big if...then statement, is needed to determine which direction the user selected. Because more than one data value needs to be passed on to the drive robot function, we need to use a merge activity to link the switch result to the three data variables. The data variables represent the speed, motor, and distance parameters. When moving the robot forward, these values will be 3, F, and 0, respectively.

The final block is an instance of the ARobotDriveService. A Join activity is used to assign the data variables to their respective variable names and then call the *DriveRobot* function from the ARobotDriveService. The final diagram should look similar to Figure 7-5.

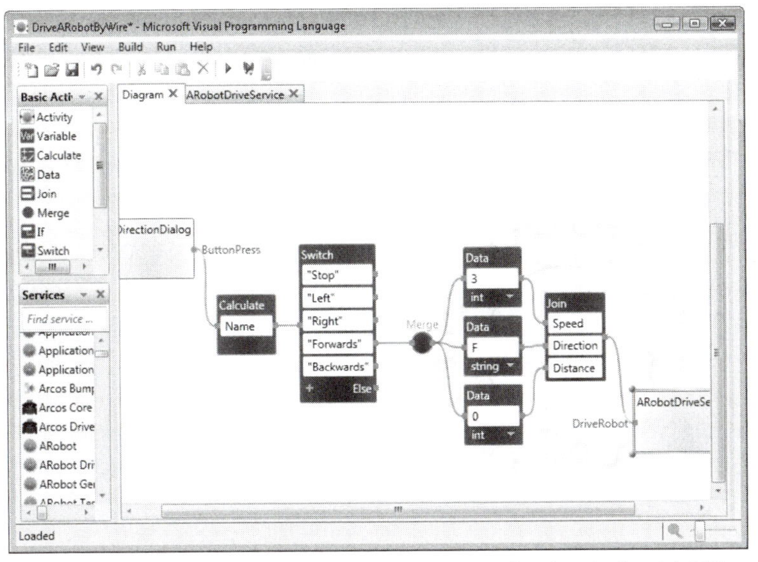

Figure 7-5 The DriveARobotByWire is a test application built with VPL and designed to test the ARobot services created in this chapter.

Summary

- In this chapter, you learned about the steps needed to interface with an unsupported robot named ARobot. ARobot is a three-wheeled robot made by Arrick Robotics.
- To support a new hardware interface, you need to create an onboard remote interface. How you create the onboard interface will vary depending on the processing requirements of the robot you are trying to control. The onboard interface for the ARobot was written using a BASIC program and downloaded to the ARobot through a serial connection.
- A core service will be used to represent the robot to MSRS and as the interface between the onboard remote interface and the high-level services. For each of the sensors and actuators supported by the ARobot, a high-level service needs to be created. For example, there will be one for the drive function and one for the whiskers.
- A Test service can be written as a regular service and included in the ARobot solution. Alternatively, the ARobot services can be tested using a VPL application, which references one of the ARobot's high-level services.

Chapter 8

Building a Security Monitor

Now it is time to build a robotics application with a useful purpose. This chapter will expand on the last chapter by using a modified ARobot, an onboard laptop, and a Web camera to create your very own security monitor robot. You or someone you know might have experience using a Web camera to monitor a home while away. Although it can be quite useful to view images from a Web camera remotely, typically the camera is static and does not move. A Web camera attached to a mobile robot can be moved around an area, thus allowing the observer to change the area being observed.

In this chapter, we will build an application that you can use to operate an ARobot both remotely and by streaming images from a Web camera attached to an onboard laptop. The application will also be used to detect motion and send an e-mail that notifies the owner of the movement. This can be very useful as a security monitor that allows owners to observe areas in their home remotely.

Working with the ARobot

The ARobot, which was introduced in Chapter 7, is a three-wheeled mobile robot available from Arrick Robotics (*http://www.robotics.com*). This expandable robot allows you to add attachments via pre-drilled holes in the metal chassis. To accommodate the application created in this chapter, you will need to add an onboard laptop (see Figure 8-1). The use of a laptop is necessary because we will be using a USB-enabled Web camera. The camera will be connected to the laptop and not the robot.

Figure 8-1 An ARobot that has been expanded by adding a laptop to the metal chassis.

Tip One alternative to using an onboard laptop is to use an embedded personal computer (PC) that runs the Compact Framework. You can purchase an embedded kit that includes all of the software and hardware you will need through the following URL: *http://www.embeddedpc.net/ebox2300MSJK/tabid/111/default.aspx*.

The cost of the kit is $250 U.S. dollars, and it includes an embedded system built on Vortex86 System-On-Chip technology, with 128 MB system memory, integrated audio, Ethernet, two serial ports, VGA, and 256 MB Integrated Drive Electronics (IDE) bootable flash storage. The Microsoft Robotics Studio (MSRS) team used this kit to create the sumo robot that was used in the Microsoft Mobile and Embedded DevCon (MEDC) 2007. Developers can get more information on how to assemble their own sumo bot from the following Web site: *http://msdn2.microsoft.com/en-us/robotics/bb403184.aspx*.

The important thing to note in this chapter is that you are not limited to working with what comes in a robotics kit. Most robot kits are expandable and allow you to add sensors or actuators. Additionally, you can assemble your own custom robot using various electronic components. For example, one Microsoft employee from the Windows Mobile development team created his own robot using a smart phone, a LEGO NXT base, and MSRS (see the sidebar titled "Profile: Brian Cross").

Profile: Brian Cross

Brian Cross, inventor of the WiMo robot.

Brian Cross has worked in the Windows Mobile group at Microsoft for several years. Cross, who earned a master's degree in computer science from the University of Washington, is responsible for developing new application programming interfaces (APIs) for developers and for enhancing the developers' experience. He earned this title because of his tenacious spirit, which was boldly demonstrated when he took it upon himself to build a robot using a Windows Mobile device.

In part, Cross built the robot he named WiMo (pronounced "Wee-mo", see *http://www.wimobot.com/*) because he has always been interested in robotics and electronic hardware. However, he specifically chose to use a Windows Mobile device because he wanted "...a fun way to show off many of the different features our phones offer in our software developer kit (SDK)." Cross also told me the following:

> *At some point, everyone will upgrade to a newer device. When someone does upgrade, what usually ends up happening to the device that they replaced? Often it sits on a shelf, gets lost, or even thrown away. What I wanted to show was that, while this device might be older, it is still quite a powerful little processor that can still be put to good use. At the very least, older devices could still be valuable in education.*

In addition to demonstrating the features in the Mobile Developer SDK, WiMo uses MSRS. Cross continues to add new features and build new versions of the mobile bot. Besides the original "home brew" version (see *http://www.wimobot.com/robots/WiMo*), there is a LEGO NXT version (see *http://www.wimobot.com/robots/NXT*). WiMo itself uses a smart phone and can optionally be controlled remotely using a PC.

Cross likes using MSRS because, after the services have been created, additional programs can be easily written using VPL. He likes the modular design and use of generic services that MSRS provides. Cross said that in addition to the use of generic services, "WiMo has a service that manages the camera of the Windows Mobile device. With that service, anyone could take their Windows Mobile device and add it to their robot (WiMo or not) and use the camera just by plugging in the WiMo Camera Service."

Cross believes that the field of robotics is exciting and only just beginning to emerge. He can see a bright future in which robots will become ubiquitous with daily life. In terms of the immediate future, he believes that:

> *With personal robotic vacuum cleaners being somewhat normal nowadays, it is only a matter of time before more of our daily tasks are automated by a robot. Telepresence is also an area that I think could grow in the coming years. The ability to "be" somewhere else without having to travel can be quite compelling, especially as the mechanics and hardware of these telepresence robots advance.*

Integrating Vision

Robotic vision is one area of robotics that has proved to be challenging. Although you can use a Web camera to simply stream images as a bitmap, this is typically not useful if you need the robot to act based on what it observes. It gets even more complicated if you need to identify and categorize specific objects, for example, if you need the robot to distinguish a cat from a dog. Even the task of simple face recognition has proved to be quite complex.

In this chapter, we will build a security monitor service that determines whether movement has occurred. This is accomplished by comparing the frame images captured from a Web camera. If there is a difference between the images, then the service will send an e-mail notification. This is one of the simplest vision tasks to accomplish, but you might need to do something more complicated. If this is the case, you might want to consider integrating robotic vision software with your service.

RoboRealm (*http://www.roborealm.com*) provides a free-to-download robotic vision software application that you can use to perform complex image analysis. The RoboRealm image tool (see Figure 8-2) can be used to perform analysis functions such as locating the center of gravity or providing color and geometric statistics. Various filters can be applied to an image that perform such functions as removing the background or finding the max, min, mean, median, and midpoints of an image. For example, you can use the minimum filter to enhance dark values in an image. The list of functions goes on and on, and you can use RoboRealm to accomplish tasks such as trail following, object location, or digital gauge reading.

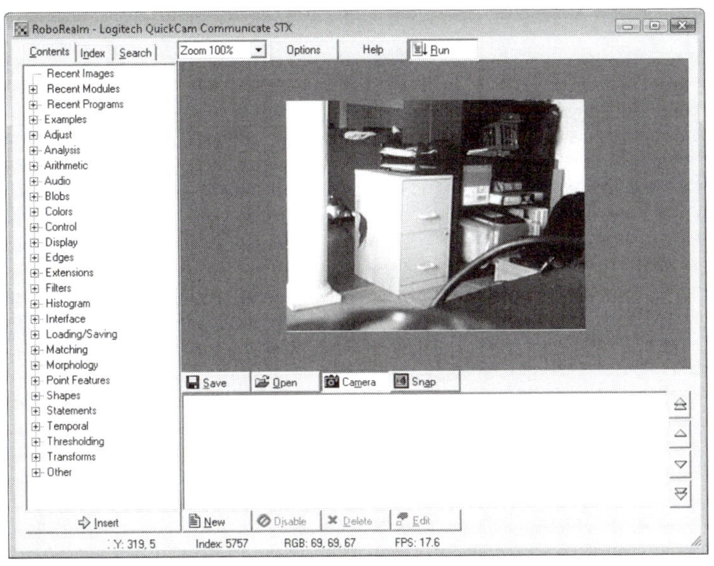

Figure 8-2 RoboRealm is a free-to-download robotic vision software tool that can be used to perform complex image analysis.

RoboRealm provides an interface service for integrating with MSRS. This allows you to use RoboRealm to process the image and then have your MSRS service instruct the robot to perform an applicable action. The MSRS interface can be downloaded from the RoboRealm Web site: *http://roborealm.com/help/MSRS.php*.

Another third-party tool that you can use to provide vision tracking services to your MSRS application is the VOLTS-IQ SDK by BrainTech (see *http://www.braintech.com*). VOLTS-IQ interfaces with the WebCam service offered by MSRS to provide real-time target tracking. You can download the latest version from the VOLTS-IQ Web site (*http://www.volts-iq.com*), and this includes sample code to get started. The software is free for noncommercial use.

Build a Security Monitor Service

The security monitor service is responsible for receiving images from the attached Web camera and determining if motion has occurred within the image. This happens by taking snapshots every few milliseconds and comparing the binary output for these images. If the service determines that motion has occurred, it will send an e-mail to a predefined e-mail address.

Users will be able to monitor images captured by the service from a Control Panel Web page (see Figure 8-3). The Control Panel Web page also includes directional buttons that you can use to navigate the ARobot through a room. Additionally, the user can enable or disable e-mail notifications and enter an e-mail address in which movement notifications should be sent.

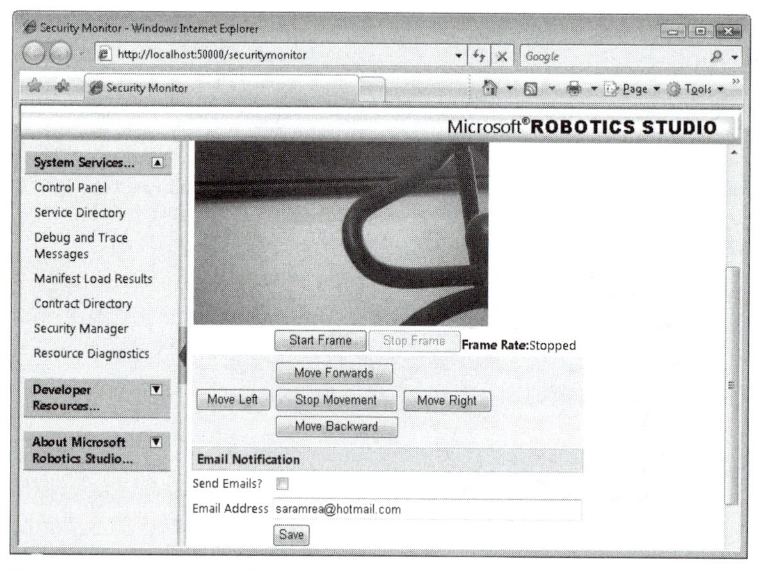

Figure 8-3 The Control Panel Web page allows the robot owner to remotely monitor and control the ARobot.

To build the security monitor service, you will need to create a new Decentralized Software Services (DSS) service using the Visual Studio template provided with MSRS. The service should be named SecurityMonitor, and it should be located in the samples folder, where the MSRS installation folders reside.

Note This chapter will not list all the code necessary to build the Security Monitor service. For readers interested in using the service, it is included on the book's companion Web site.

Defining the Service Contract

The security monitor service requires several state variables. You define these variables in the contract class, which you should name SecurityMonitorTypes.cs. You can define the state variables with the following code:

```
[DataContract()]
public class SecurityMonitorState
{
    [Description("Specifies width of frame displayed in Control Panel")]
    public const int ImageWidth = 320;

    [Description("Specifies height of frame displayed in Control Panel")]
    public const int ImageHeight = 240;
```

```
[Description("Indicates refresh interval for webcam service (in ms).")]
public const int RefreshInterval = 250;

private Bitmap _image = new Bitmap(320, 240);
public Bitmap Image
{
    get { return _image; }
    set { _image = value; }
}

private Bitmap _motionImage = new Bitmap(320, 240);
public Bitmap motionImage
{
    get { return _motionImage; }
    set { _motionImage = value; }
}


[DataMember]
[Description("Size of the image stored in state.")]
public Size Size
{
    get
    {
       if (_image != null)
       {
          return _image.Size;
       }
       return Size.Empty;
    }
    set { return; }
}

private DateTime _timeStamp;
[DataMember]
[Description("Time that the image was retrieved.")]
public DateTime TimeStamp
{
   get { return _timeStamp; }
   set { _timeStamp = value; }
}

private Boolean _sendEmail;
[DataMember]
[Description("Indicates whether we send emails sent to the address.")]
public Boolean SendEmail
{
   get { return _sendEmail; }
   set { _sendEmail = value; }
}

private String _emailAddress;
[DataMember]
[Description("Email address to send notifications to.")]
```

```
    public String EmailAddress
    {
        get { return _emailAddress; }
        set { _emailAddress = value; }
    }

    public ObjectResult CurObjectResult;

}
```

Modify the PortSet

You will need to modify the PortSet to include the *HttpGet*, *HttpPost*, *HttpQuery*, *Replace*, and *NotifyMotionDetection* operations. The *Http* operations will be used when we send and receive state variable values from the Control Panel Web page. The PortSet also includes operations used to drive the ARobot. The code for the security monitor main operations port should look like the following:

```
[ServicePort()]
public class SecurityMonitorOperations : PortSet<DsspDefaultLookup,
     DsspDefaultDrop,
     Get,
     HttpGet,
     HttpPost,
     HttpQuery,
     Replace,
     NotifyMotionDetection>
{
}
```

Add New Service Operations

Because the *Http* operations are already supported by MSRS, you do not need to add code to the SecurityMonitorTypes.cs file to accommodate the operations. Instead, you need to add a reference to the *DSSP HTTP* namespace. You can do so by adding the following code to the top of the file:

```
using Microsoft.Dss.Core.DsspHttp;
```

You need to add code to support the *NotifyMotionDetection* operation because this is a custom operation not provided with MSRS. Add the following code to support the new operation below the PortSet:

```
[DisplayName("NotifyMotionDetection")]
[Description("Indicates that a moving object has been detected.")]
public class NotifyMotionDetection : Update<ObjectResult,
      PortSet<DefaultUpdateResponseType, Fault>>
{
    public NotifyMotionDetection()
    {
```

```
    }
    public NotifyMotionDetection(ObjectResult body) : base(body)
    {
    }
}
```

Add Code to the Implementation Class

The implementation class will be named SecurityMonitor.cs, and this is where the code used to monitor and control your robot will reside. For the security monitor service, you will need to add references to the following four assemblies:

- ***RoboticsCommon.Proxy*** This assembly, which is included with MSRS, allows you to access a variety of namespaces used to control a robot. In this chapter, we will add a reference to the *Microsoft.Robotics.Services.Webcam.Proxy* namespace. This provides access to images captured by the attached Web camera.
- ***System.Drawing*** This assembly provides access to the *System.Drawing* and *System.Drawing.Images* namespaces. This provides access to functions and types that are used when processing the images captured by the Web camera.
- ***System.Web*** This assembly provides access to the *System.Web* and *System.Web.Mail* namespaces and are used when we send e-mail via Simple Mail Transfer Protocol (SMTP). It is also used when we reference the HTTP functions that stream images to the Control Panel Web page.
- ***ARobotServices.Y2007.M10.Proxy*** This assembly was created in Chapter 7 and is used to access the *Microsoft.Robotics.Services.ARobot.Drive.Proxy* namespace. This provides access to the ARobot by Arrick Robotics.

Add Partnerships

For the security monitor service, you need to add a partnership for the Webcam service and the ARobotServices service. This is how we access images from the attached Web camera and also drive the ARobot. The code for the partnership declaration should look like the following:

```
[Partner("Webcam", Contract = webcam.Contract.Identifier,
      CreationPolicy = PartnerCreationPolicy.UseExistingOrCreate)]
webcam.WebCamOperations _webCamPort = new webcam.WebCamOperations();

[Partner("ARobotDrive", Contract = drive.Contract.Identifier,
     CreationPolicy = PartnerCreationPolicy.UseExistingOrCreate,
     Optional = false)]
private drive.DriveRobotOperations _drivePort = new
     drive.DriveRobotOperations();
```

In addition to the partnership, you need to include declarations for additional ports, as well as the main port. The first port named _internalPort is used to post internal state variables. The port named _utilitiesPort is used to handle all HTTP requests and responses. The code used to declare these ports should look like the following:

```
//Used to post internal state
private SecurityMonitorOperations _internalPort = new
     SecurityMonitorOperations();

//Used to post our HTTP request responses
DsspHttpUtilitiesPort _utilitiesPort = new DsspHttpUtilitiesPort();
```

Modify the *Start* Method

The *Start* method is responsible for initializing state variables and registering the message handlers that the service uses. For this service, it is also responsible for creating an instance of the HTTP utilities port. This is the port where we send all HTTP requests and responses.

The last thing the *Start* method does is call the function *GetFrame*. This function is where we capture images from the Web camera. The code for the *Start* method (along with the associated method stubs) should look like the following:

```
protected override void Start()
{

      base.Start();

      //Initialize variables used when sending email notifications
      _state.EmailAddress = _toEmail;
      _state.SendEmail = false;

      //Initialize the Utility port used to post HTTP operations
      _utilitiesPort = DsspHttpUtilitiesService.Create(Environment);

      //Register our message handlers
      MainPortInterleave.CombineWith(
          Arbiter.Interleave(
             new TeardownReceiverGroup(),
             new ExclusiveReceiverGroup(
                Arbiter.Receive<NotifyMotionDetection>(true, _internalPort,
                    MotionDetectionHandler
                   ),
             new ConcurrentReceiverGroup(
)
        ));

        //Call the function which gets the image from the webcam
        Activate(Arbiter.ReceiveWithIterator(false, TimeoutPort(3000),
                    GetFrame));
}
public void MotionDetectionHandler(NotifyMotionDetection update){}
[ServiceHandler(ServiceHandlerBehavior.Exclusive)]
public void ReplaceHandler(Replace replace)
{
```

```
    _state = replace.Body;
    replace.ResponsePort.Post(DefaultReplaceResponseType.Instance);
}

[ServiceHandler(ServiceHandlerBehavior.Concurrent)]
public IEnumerator<ITask> HttpQueryHandler(HttpQuery query)
{
    return HttpHandler(query.Body.Context, query.ResponsePort);
}

[ServiceHandler(ServiceHandlerBehavior.Concurrent)]
public IEnumerator<ITask> HttpGetHandler(HttpGet get)
{
     return HttpHandler(get.Body.Context, get.ResponsePort);
}
IEnumerator<ITask> HttpHandler(HttpListenerContext context,
     PortSet<HttpResponseType, Fault> responsePort)
{
    yield break;
}
[ServiceHandler(ServiceHandlerBehavior.Exclusive)]
public IEnumerator<ITask> HttpPostHandler(HttpPost post)
{
    yield break;
}
public IEnumerator<ITask> GetFrame(DateTime timeout)
{
    yield break;
}
```

Note This code block also contains method stubs to represent the message handlers. For now, the body of most of these handlers is empty, and we will cover what code goes in these handlers in the next section.

You might notice that some of the handlers (such as the *Replace* and *Get* handlers) do not use an iterator. This might be different from other sample code you have looked at. The general rule here is that you should use an iterator only when there is a need to perform a yield return.

Get Images from the Web Camera

In this service, we use the Web camera service provided with MSRS to retrieve images from the attached Web camera. The *QueryFrame* method, which is one of the Web camera operations, can be used to return the most recently captured image. After the image is retrieved, we scale it to fit the image size specified in our state variables. We then call the *ProcessFrameHandler* method, which is where the image processing takes place. The code for this method is covered in the next section.

If the *ProcessFrameHandler* method determines that movement has occurred, then a notification is triggered. Regardless of whether movement is detected, the image is saved to the state variable named *Image*. The last thing the *GetFrame* function does is reset the call to itself. The code for the *GetFrame* function (along with the associated functions) is shown as follows:

```
public IEnumerator<ITask> GetFrame(DateTime timeout)
{
     //Create first query request
     //The frame returned from this request will be used
     //for processing
     webcam.QueryFrameRequest processRequest = new
           webcam.QueryFrameRequest();
     processRequest.Format = Guid.Empty;
     webcam.QueryFrameResponse frame = null;
     Fault fault = null;

     //Get the most recently captured image so it can be processed
     yield return Arbiter.Choice(
        _webCamPort.QueryFrame(processRequest),
        delegate(webcam.QueryFrameResponse success)
        {
            frame = success;
        },
            delegate(Fault f)
        {
            fault = f;
        }
     );

     //Check to see if there was a problem
     if (fault != null)
     {
        LogError(LogGroups.Console, "Could not query webcam frame", fault);
        yield break;
     }

     //If we have a valid image frame then we need to process
     //it to see if there are any moving images
     if (frame != null && frame.Frame != null)
     {
         try
         {
            //Get the current frame
            byte[] _curFrame = frame.Frame;
            int curWidth = frame.Size.Width;
            int curHeight = frame.Size.Height;
            //Set the state variables
            _state.TimeStamp = frame.TimeStamp;
            _state.Size = frame.Size;

            //See if the image needs to be scaled
            //If this is not done then we can receive
            //an error when trying to process the image
            if (_curFrame.Length != _processFrame.Length)
            {
                bool success = Scale(3, _curFrame, new RectangleType(0, 0,
                  curWidth - 1, curHeight - 1), _processFrame, ImageWidth,
```

```
                    ImageHeight);
                if (success == false)
                {
                    LogError(LogGroups.Console, "Unable to scale image");
                    yield break;
                }
                else
                {
                    _curFrame = _processFrame;
                }
            }

            // This is where actual image processing methods are processed
            ProcessFrameHandler(_processFrame, _lastProcessingResult);

            //See if movement was detected ;if so, post a notification
            if (_lastProcessingResult.MotionFound && !_firstFrame)
            {
                _state.motionImage = SaveImage(_curFrame, ImageWidth,
ImageHeight);
                _internalPort.Post(new NotifyMotionDetection(new
                  ObjectResult(_lastProcessingResult)));
            }

            //See if this is the first frame since motion will be detected
            //on that one and we want to ignore that
            if (_firstFrame) _firstFrame = false;

            //Save the image to the Image state variable
            _state.Image = SaveImage(_curFrame, ImageWidth,
ImageHeight);

           //Reset the call for the function so it will be called
           //again according to time set for the Refresh Interval.
   //This is what creates the loop
           Activate(Arbiter.ReceiveWithIterator(false,
                    TimeoutPort(RefreshInterval), GetFrame));
        }
        catch (Exception e)
        {
            LogError(LogGroups.Console, "Excp in Image Processing", e);
        }

    }
    }
    public bool Scale(int pixelBytes, byte[] sourceImage, int srcWidth,
         int srcHeight, byte[] tarImage,
         int tarWidth, int tarHeight)
    {
         if (sourceImage == null ||
           tarImage == null ||
           tarImage.Length != tarWidth * tarHeight * 3)
         {
            return false;
```

```
            }

            using (Bitmap bmp = SaveImage(sourceImage, srcWidth, srcHeight))
            {
               using (Bitmap scaled = new Bitmap(bmp, tarWidth, tarHeight))
               {
                  BitmapData locked = scaled.LockBits(
                      new Rectangle(0, 0, tarWidth, tarHeight),
                      ImageLockMode.ReadOnly,
                      PixelFormat.Format24bppRgb
                  );

                  Marshal.Copy(locked.Scan0, tarImage, 0, tarImage.Length);

                  scaled.UnlockBits(locked);
               }
           }
               return true;
       }

public Bitmap SaveImage(byte[] curFrame, int width, int height)
{
     //Create a bitmap image variable
     //which will hold a bitmap
     //representation of the current frame
     Bitmap bmp = new Bitmap(
         width,
         height,
         PixelFormat.Format24bppRgb
     );

     // Use the LockBits function to lock the image
     //into memory
     BitmapData data = bmp.LockBits(
         new Rectangle(0, 0, width, height),
         ImageLockMode.ReadOnly,
         PixelFormat.Format24bppRgb
     );
     Marshal.Copy(curFrame, 0, data.Scan0, curFrame.Length);

     //Unlock the image from system memory and store it
     //in the bitmap image variable
     bmp.UnlockBits(data);

     //Pass back the bitmap image
     //which will be stored in a state variable
     return bmp;
}
public void ProcessFrameHandler(byte[] rgbValues, ImageProcessingResult
   result) {}
```

Tip Keep in mind that modifying the state, such as when the image is saved to an image state variable, can cause a potential conflict. The conflict could occur when the *Get* or *Replace* function is executing. Even though the likelihood of this is low, it should still be considered when designing your own service.

Note This code block also contains a method stub to represent the *ProcessFrameHandler* method because this we cover this method in the next section.

Detect Motion

The *ProcessFrameHandler* method is where we process each frame captured by the Web camera. The frame image is stored in a byte array and is represented through a series of binary values. Each binary value represents a single pixel. The size of the array is determined by the state variables that represent the image's height and width. The code for the *ProcessFrameHandler* method, along with the *DetectMotion* function that it calls, are seen as follows:

```
byte[] _processFrame = new byte[ImageWidth * ImageHeight * 3];
byte[] _grayImg = new byte[ImageWidth * ImageHeight];
byte[] _grayImgOld = new byte[ImageWidth * ImageHeight];
byte[] _motionImg = new byte[ImageWidth * ImageHeight];

public void ProcessFrameHandler(byte[] rgbValues, ImageProcessingResult
    result)
{
    if (rgbValues == null) return;

    //Strip out the red, green, and blue colors and make the image gray
    for (int i = 0, j = 0; i < _grayImg.Length; i++, j += 3)
    {
       _grayImg[i] = (byte)(((int)rgbValues[j]
                 + (int)rgbValues[j + 1]
                 + (int)rgbValues[j + 2]) / 3);
    }

    //zero out the motion image array
    for (int i = 0; i < _motionImg.Length; i++)
    {
       _motionImg[i] = 0;
    }

    // calculate frame difference by computing the
    //difference between the current image and the
    //last captured image
    for (int i = 0; i < ImageWidth * ImageHeight; i++)
    {
       _motionImg[i] = (byte)Math.Abs(_grayImg[i] - _grayImgOld[i]);
```

```
    }

    //see if there is motion by calling the DetectMotion function
    bool motionFound = DetectMotion(_motionImg, 40, 250, result);

    //Copy the current image to the _grayImgOld array so it
    //can be compared to the next frame captured
    _grayImg.CopyTo(_grayImgOld, 0);
}

static bool DetectMotion(byte[] image, byte diffThresh, int sizeThresh,
     ImageProcessingResult result)
{
      int i, x, y;

      //Compare the values in the motion image against
      //the allowed difference threshhold value of 40
      int sumX = 0;
      int sumY = 0;
      int pixelCount = 0;
      for (i = 0; i < ImageWidth * ImageHeight; i++)
      {
          if (image[i] > diffThresh)
          {
              image[i] = 255;
              sumX += i % ImageWidth;
              sumY += i / ImageWidth;
              pixelCount++;
          }
          else
          {
              image[i] = 0;
          }
      }


      //See if the number of pixel differences
      //exceeds the size threshhold value of 250
      //which indicates that motion was detected
      if (pixelCount > sizeThresh)
      {
           result.MotionFound = true;
           result.XMotion = (sumX / pixelCount);
           result.YMotion = (sumY / pixelCount);
           result.MotionSize = pixelCount;
      }
      else
           result.MotionFound = false;

      //Pass back the result
      return result.MotionFound;
}
```

The first thing the *ProcessFrameHandler* method does is strip out the red, green, and blue colors to create a monochrome or gray image. The gray image is much smaller in size and, therefore, quicker to process. The monochrome image is stored in a byte array variable named *grayImg*.

To determine if motion has occurred, we need to compare the binary values in the *grayImg* array to values in another array named *grayImgOld*. The *grayImgOld* array will hold binary values from the last frame captured. The observant reader should note that the first time *ProcessFrameHandler* is called, the values stored in *grayImgOld* will be zero. This is because we do not copy values into this array until the end of the *ProcessFrameHandler* method. Therefore, the first time this method is called, it detects motion, even though it has not really occurred.

To prevent the *GetFrame* function from triggering an invalid notification, we use a variable named _firstFrame. This variable is initialized with a value of true. The code that immediately follows the call to *ProcessFrameHandler* checks to see if motion was found. It also checks to see if the _firstFrame variable is false. Only when the variable is false will the motion notification be triggered. After the first frame is processed, the variable is set with a value of false, thus allowing normal motion checking to occur.

Motion is detected when the difference between the current frame and the last frame exceeds a threshold value. For our service, the allowed threshold difference is a binary value of 40. The *DetectMotion* function accepts as an input parameter the array named *_motionImg*. This array was populated in the *ProcessFrameHandler* method, and it contains the calculated difference between the current frame and the last frame. *DetectMotion* is responsible for looping through the *_motionImg* array and counting the number of times that the difference exceeds the value of 40. If the number of pixels that exceeded the threshold is greater than 250, then we determine that motion has occurred.

Send an E-Mail Notification

When the service detects motion, it triggers an e-mail notification by posting data to the internal port for the *NotifyMotionDetection* operation. This is done in the *GetFrame* method after the *ProcessFrameHandler* method is called. A copy of the image is also saved to a state variable named *MotionImage*. The portion of code where this occurs is shown as follows:

```
//See if movement was detected ;if so, post a notification
if (_lastProcessingResult.MotionFound && !_firstFrame)
{
    if (_state.motionImage != null)
    {
        _state.motionImage.Dispose();
    }
    _state.motionImage = new Bitmap(_state.Image);
     _internalPort.Post(new NotifyMotionDetection(new
        ObjectResult(_lastProcessingResult)));
}
```

The observant reader might wonder why we are saving the image twice because we also save a copy of the image in a state variable named *Image*. The reason two copies of the image are saved is because the *Image* state variable is overwritten the next time the *GetFrame* method is called. We also dispose of the original motion image when it is no longer needed.

After the data is posted to the internal port, it triggers the message handler named *MotionDetectionHandler* to fire. This handler is responsible for sending an e-mail to the address specified in the state variable named *EmailAddress*. It only does this if the state variable named *SendEmail* is set with a value of true. The code for the *MotionDetectionHandler* method is shown as follows:

```
public void MotionDetectionHandler(NotifyMotionDetection update)
{
     _state.CurObjectResult = update.Body;

     //First see if we should send notifications
     if (_state.SendEmail)
     {
         try
         {
             //If the email notification is enabled, send an
             //email to the appropriate parties
             StringBuilder sb = new StringBuilder();
             sb.Append("Motion was detected by your at-home Security ");
             sb.Append("Monitor robot. Go to the link below to see");
             sb.Append("what the motion image looked like.<br>");
             sb.Append("<a target=_blank href=");
 sb.Append(_publicURL + ">Motion Image</a>");

             string subject = "Motion Detected";
             MailMessage message = new MailMessage();
             message.From = _fromEmail;
             message.To = _state.EmailAddress;
             message.Subject = subject;
             message.Body = sb.ToString();
             message.BodyFormat = MailFormat.Html;
             SmtpMail.SmtpServer = _hostName;
             SmtpMail.Send(message);

         }
         catch (Exception e)
         {
             LogError(LogGroups.Console, "Exception in sending email notification", e);
         }
         finally
         {
             //Turn off the email notifications so that we do not
             //keep receiving emails. The end user will have to
             //re-enable the emails from Control Panel after they
             //have viewed the image
             _state.SendEmail = false;
         }
```

```
    }

    //Post the response
    update.ResponsePort.Post(DefaultUpdateResponseType.Instance);
}
```

For the *MotionDetectionHandler* to successfully send an e-mail, SMTP must be enabled and running on the machine that is operating the service. For the security monitor service, this is the laptop machine that is located on top of the ARobot.

Note By default, the SMTP service is not installed on nonserver machines such as Windows XP or Windows Vista. You will need to refer to the documentation for your operating system to determine how to install and configure SMTP. After it's installed, SMTP can be accessed through Control Panel and Services.

Create the Transformation File

To display the state variables in a Web browser, we use an Extensible Stylesheet Language Transformation (XSLT) file. Using an XSLT file allows you to programmatically control how the state is displayed in a Web browser. It is preferable to display the state in a Web browser because the Web page can be accessed remotely. For example, a robot owner can access the service Web page from his or her office, even though the robot is located at home.

MSRS includes a DSS XSLT template that you can use to give your service a consistent look and feel. This means that the HTML rendered by the XSLT file will be surrounded by the same interface shell used to access the MSRS Web interface (see Figure 8-4).

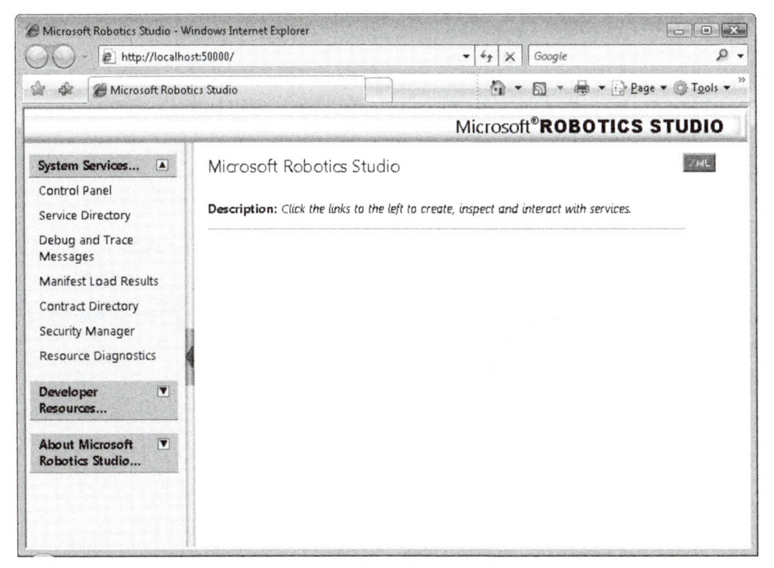

Figure 8-4 MSRS interface, which is displayed when you run the DSS node.

To use the template, you first need to add a reference to the transform URL. This is done at the top of the XSLT file. For example, the following code should appear in the file named ControlPanel.xslt:

```
<?xml version="1.0" encoding="UTF-8" ?>
<xsl:stylesheet version="1.0"
  xmlns:xsl="http://www.w3.org/1999/XSL/Transform"
  xmlns:wc="http://schemas.tempuri.org/2007/11/securitymonitor.html">
```

In this code, we assign the xsl prefix to the transform URL. This means that every time we need to use transform operatives, they will include the xsl prefix. We also include a reference to the security monitor service by including the service contract URL. In this case, we use the wc prefix. This means that every time we need to reference state variables in the security monitor service, we will include the wc prefix.

The next thing to do is import the master page used by the DSS template. This is done with the following line of code:

```
<xsl:import
  href="/resources/dss/Microsoft.Dss.Runtime.Home.MasterPage.xslt" />
```

By importing the master page, we do not have to include HTML tags such as <head> and <body>. Instead we will use code such as the following to override the master page parameters:

```
<xsl:template match="/">
   <xsl:call-template name="MasterPage">
     <xsl:with-param name="serviceName">
       Security Monitor
     </xsl:with-param>
     <xsl:with-param name="description">
       Control Panel
     </xsl:with-param>
     <xsl:with-param name="head">
```

The next portion of the XSLT file will contain JavaScript. The JavaScript is used to manipulate values passed back in the service state. It is also used to respond to Web page events such as loading the page or button clicks.

The remainder of the XSLT file contains the HTML and XSL operations that will be used to render the control panel Web page. A <form> tag is included because we will post back the results to the security monitor service. The remainder of the XSLT file will look like the following:

```
<xsl:template match="/wc:SecurityMonitorState">

   <form id="oForm" action="" method="post">
      <table border="0">
         <tr class="odd">
           <th colspan="2">
              Camera
           </th>
         </tr>
```

```
<tr class="even">
  <td colspan="2" align="left">
    <img id="TargetImg" name="TargetImg"
        src="/securitymonitor/jpeg"
        alt="Camera Image"
        onload="onImageLoad()"
        onerror="onImageError()">
      <xsl:attribute name="width">
        <xsl:value-of select="wc:Size/wc:Width"/>
      </xsl:attribute>
      <xsl:attribute name="height">
        <xsl:value-of select="wc:Size/wc:Height"/>
      </xsl:attribute>
    </img>
  </td>
</tr>
<tr class="even">
  <td></td>
  <td>
    <button id="btnStart" name="btnStart" onclick="startFeed()">
        Start</button>
    <button id="btnStop" name="btnStop" onclick="stopFeed()"
          disabled="disabled">
        Stop</button>
    Frame Rate
    <span id="spanFrameRate">Stopped</span>
  </td>
</tr>
<tr class="odd">
  <td colspan="2">
     <table>
       <tr>
         <td></td>
         <td>
           <button id="btnForward" name="btnForward"
               onclick="moveForward()">
             Move Forwards
           </button>
         </td>
         <td></td>
       </tr>
       <tr>
         <td>
           <button id="btnLeft" name="btnLeft"
onclick="moveLeft()">
             Move Left
           </button>
         </td>
         <td>
           <button id="btnStop" name="btnStop"
onclick="stopMovement()">
             Stop Movement
           </button>
         </td>
         <td>
```

```
                    <button id="btnRight" name="btnRight"
onclick="moveRight()">
                      Move Right
                    </button>
                  </td>
                </tr>
                <tr>
                  <td></td>
                  <td>
                    <button id="btnBackward" name="btnBackward"
onclick="moveBackward()">
                      Move Backward
                    </button>
                  </td>
                  <td></td>
                </tr>
              </table>
            </td>
          </tr>
         <tr class="even">
           <th colspan="2">
             Email Notification
           </th>
         </tr>
         <tr>
           <td>
             Send Emails?
           </td>
           <xsl:if test="wc:SendEmail = 'true'">
             <td>
               <input type="checkbox"
                    class="storeuserData"
                    checked="checked"
                    name="sendEmail" />
             </td>
           </xsl:if>
           <xsl:if test="wc:SendEmail = 'false'">
             <td>
               <input type="checkbox"
                      class="storeuserData"
                      name="sendEmail" />
             </td>
           </xsl:if>
         </tr>
         <tr>
           <td>
              Email Address
           </td>
           <td>
             <input type="Text"
                    class="storeuserData"
                    name="emailAddress"
                    size="50"/>
             <xsl:attribute name="value">
               <xsl:value-of select="wc:EmailAddress"/>
```

```
              </xsl:attribute>
            </td>
          </tr>
          <tr class="even">
            <td></td>
            <td>
              <input type="submit" name="Save" Value="Save"
                 onclick="saveEmail()"/>
            </td>
          </tr>
        </table>
      </form>

</xsl:template>
```

Each of the buttons on the Control Panel Web page is associated with a click event. The click event refers to the name of a JavaScript function, which is included at the top of the XSLT file. For example, when the user clicks the Left button, the *moveLeft* function (shown below) is called:

```
function moveLeft()
{
    saveInput();
}
```

The *moveLeft* function performs one action: it calls the *saveInput* function, which is shown as follows:

```
function saveInput()
{
   if (fLoaded)
   {
       var oPersist = document.all(sPersistObject);
       oPersist.setAttribute("sInterval", oPersist.value);
       oPersist.setAttribute("sRunning", feedRunning);
       oPersist.save(sStore);
   }
}
```

The *saveInput* function in turn triggers a postback and causes the *HttpPostHandler* to execute. The code for the *HttpPostHandler* reads the form data, and, based on the data posted, it moves the robot in the appropriate direction. The code for the *HttpPostHandler* is shown as follows:

```
[ServiceHandler(ServiceHandlerBehavior.Exclusive)]
public IEnumerator<ITask> HttpPostHandler(HttpPost post)
{
     Fault fault = null;
     NameValueCollection collection = null;

     //Read the data on the Web page form
     ReadFormData readForm = new ReadFormData(post.Body.Context);
    _utilitiesPort.Post(readForm);
```

```
// Get the data on the Web page and store it in a collection variable
yield return Arbiter.Choice(
    readForm.ResultPort,
   delegate(NameValueCollection col)
   {
      collection = col;
   },
   delegate(Exception e)
   {
      fault = Fault.FromException(e);
      LogError(null, "Error processing form data", fault);
   }
 );

//Check for an error
if (fault != null)
{
    post.ResponsePort.Post(fault);
    yield break;
}
if (!string.IsNullOrEmpty(collection["Forward"]))
{
    //Send a command to drive forward for no specific distance
   _drivePort.DriveRobot(new drive.DriveRobotRequest(6, "F", 0));
}

if (!string.IsNullOrEmpty(collection["Backward"]))
{
    //Send a command to drive backward for no specific distance
   _drivePort.DriveRobot(new drive.DriveRobotRequest(6, "B", 0));
}

if (!string.IsNullOrEmpty(collection["Left"]))
{
    //Send a command to drive Left for no specific distance
   _drivePort.DriveRobot(new drive.DriveRobotRequest(6, "L", 0));
}

if (!string.IsNullOrEmpty(collection["Right"]))
{
    //Send a command to drive Right for no specific distance
   _drivePort.DriveRobot(new drive.DriveRobotRequest(6, "R", 0));
}

if (!string.IsNullOrEmpty(collection["Stop"]))
{
    //Send a command to stop
   _drivePort.DriveRobot(new drive.DriveRobotRequest(6, "S", 0));
}

//Check for an error again
if (fault != null)
{
    post.ResponsePort.Post(fault);
    yield break;
```

```
    }

    //Return response saying everything was ok
    post.ResponsePort.Post(new HttpResponseType(HttpStatusCode.OK, _
        state, _transform));
    yield break;
}
```

Create a Manifest File

You can create the manifest file for the security monitor service by using the DSS Manifest Editor. Before you can do so, you must successfully compile the security monitor service because this creates the assemblies needed by the Manifest Editor.

Tip When you open the DSS Manifest Editor, it caches a copy of each service assembly. If you make a change to your service that involves a partner service, you need to exit and re-enter the Manifest Editor before the change will be recognized.

To begin, open the Manifest Editor and drag an instance of the SecurityMonitor service onto the design workspace. You then need to drag an instance of the WebCam service onto the node named *Webcam* and an instance of the ARobotDriveService service onto the node named *ARobotDrive*. When complete, the manifest should look like figure 8-5.

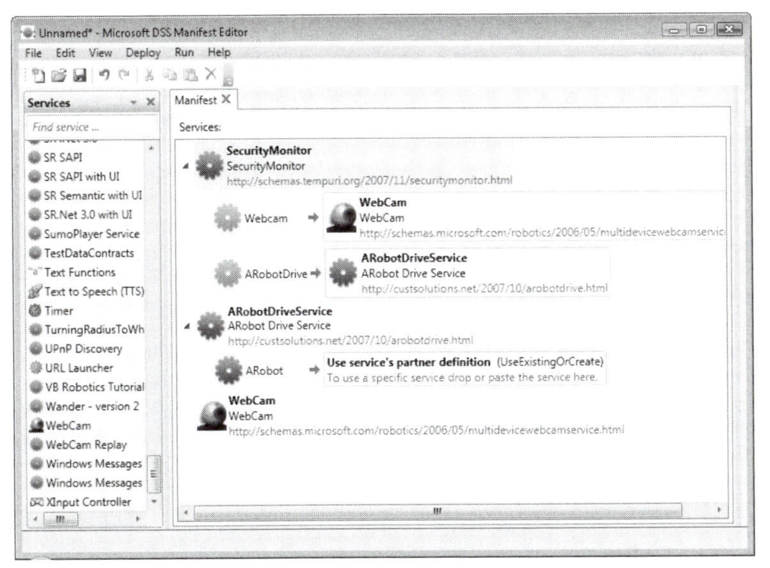

Figure 8-5 Complete manifest for the security monitor service.

You can then save the complete manifest to the code directory for the security monitor service. At this point, you are ready to run the service and obtain observations from the Web camera.

Run the Security Monitor Service

Before mounting your laptop onto the ARobot and attempting to operate it using the service, you may want to ensure that the Web camera is operating successfully. When you start the service, you should wait until the service completes loading each manifest in the command window before opening a Web browser. You can then open a Web browser and enter the following URL to access the control panel Web page: *http://localhost:50000/securitymonitor*.

By accessing this URL, you see the full rendering of the XSLT file. Optionally, you can append "/jpeg" to the end of the URL to render the latest image captured from the Web camera. Additionally, you can see the image captured when motion occurs by entering the following URL: *http://localhost:50000/securitymonitor/motion*.

Summary

- In this chapter, we built a security monitor service by modifying the ARobot and attaching an onboard laptop with a Web camera. We use this service to capture images from the attached Web camera and display them on a Control Panel Web page. You can also use the Control Panel to control the robot by moving it left, right, forward, backward, or stopping completely.
- The security monitor service partners with the Web camera service to capture images from the attached Web camera. It also partners with the ARobot services created in Chapter 7. These services are used to drive the ARobot using directional buttons on the control panel Web page.
- Images are captured from the Web camera every 250 milliseconds, and the result is stored in an array of binary values. The motion detection portion of the service simply compares the binary values from the current frame with the last frame captured to determine if motion has occurred. If it has, then it triggers a notification and sends an e-mail to the e-mail address stored in one of the state variables.
- An XSLT file named ControlPanel.xslt is used to render the service state as HTML. This also allows the robot owner to access the control panel Web application remotely.

Chapter 9

The Future of Robotics

Robotics has on occasion fallen victim to overly optimistic and unrealistic expectations. Many of the reasons for this failure to accurately predict its future are because the field is extremely complex. Even though some of the limitations involving mechatronics have been addressed, there are still many software problems left to tackle. This chapter will limit predictions and specific information to a time period within the next five years. It will also cover areas that are well established or applications already in development.

Future of MSRS

The current version of Microsoft Robotics Studio (MSRS) is 1.5. Sometime in the first half of 2008, the MSRS team expects to release a Community Technology Preview (CTP) for version 2.0. The CTP is the first step toward releasing the final version, and the team will look for feedback from the development community before the final release.

The 2.0 version should provide significant enhancements to the 1.5 version. Most notably, there will be big improvements to VPL that will allow it to be useful when programming embedded devices across the network and across various embedded operating systems (OS). In fact, the 2.0 version will prime the way for MSRS to be tightly integrated with Windows Mobile devices.

In late 2007, the Mobile and Embedded Devices (MED) division at Microsoft (see *http://msdn2.microsoft.com/embedded/default.aspx*) adopted Concurrency and Coordination Runtime (CCR), Decentralized Software Services (DSS), and VPL as its distributed programming model. The announcement was made at the September 2007 Embedded Systems Conference in Boston, and you can obtain more information from the following transcript: *www.windowsfordevices.com/files/misc/Kevin_Dallas_ESC_East2007.pdf*. Distributed embedded systems will have a great impact on developers of sensor networks and factory automation applications.

Throughout the writing of this book, I have been fortunate enough to have several members of the MSRS team review each chapter for technical accuracy. George Chrysanthakopoulos (refer to the sidebar titled "Profile: George Chrysanthakopoulos and Henrik F. Nielsen") recently responded to my question about whether MSRS would be integrated with other product offerings with the following:

> *...the technologies in MSRS are already being seen as critical for a variety of non-robotic applications, ranging from sensor networks and industrial automation to managing concurrency on high-end systems. There are plans to integrate the CCR with the mainstream Microsoft concurrency effort, the Parallel Computing Framework, as well.*

Profile: George Chrysanthakopoulos and Henrik F. Nielsen

George Chrysanthakopoulos (left) and Henrik Nielsen are principal members of the MSRS team and inventors of the CCR and DSS.

George Chrysanthakopoulos and Henrik F. Nielsen are two of just a handful of developers on the MSRS team. Even though the team is small, the experience and proficiency of its members make it a highly productive group. Chrysanthakopoulos, who holds a PhD in electrical engineering from the University of Washington, and Nielsen, who received a master's degree in engineering from the Aalborg University and worked at CERN, are responsible for designing the key technologies that drive MSRS – the DSS and CCR models.

Nielsen, who has worked with Tim Berners-Lee (inventor of the World Wide Web) and Roy Fielding (designer of the Representational State Transfer, or REST, model), is one of the principal architects for several of the HTTP and SOAP specifications. Before joining Microsoft in 1999, he worked on the technical staff at the World Wide Web Consortium (W3C). Chrysanthakopoulos, who has worked in several groups at Microsoft, joined the Advanced Technologies and Strategy group in 2003. It was there that he met Nielsen, and the two of them realized they had a similar vision for creating a Web-based distribution model based on a state-driven approach.

During a two-year period, Chrysanthakopoulos and Nielsen worked on both the CCR and DSS. DSS uses the CCR to manage the enormous amount of asynchronous processing that a distributed model requires. The work that Chrysanthakopoulos and Nielsen did eventually led them to the MSRS group because DSS and the CCR seemed the perfect foundation for robotics applications. Chrysanthakopoulos currently holds the title of Software Architect and Development Manager, while Nielsen serves as the team's Group Program Manager.

Potential Applications

In this section, we examine viable and practical applications of robotics. This involves potential applications that are either under development or in the process of maturing. You should note that all of these applications have specific purposes. General purpose intelligence is an area of robotics and artificial intelligence (AI) still under development.

Note The applications presented in this section do not necessarily use MSRS as their software platform. However, there is no reason that they could not use MSRS in the future.

Smart Appliances

Since 1994, Microsoft researchers have maintained an area inside the Executive Briefing Center that is known as the Microsoft Home. This home of the future is updated every few years with the latest in gadgets and smart appliances. Microsoft hopes these devices will be used in most homes within the next 5 to 10 years. The simulated home, which does not include laptops or desktop machines, uses mesh networking and thin LCD screens to interact with users.

Microsoft Home has become a test bed for new technologies within Microsoft. It represents an effort to make the family home more efficient and enjoyable. Products are embedded within the home, such as the digital wallpaper that routinely changes the images displayed on the wall. For more information about Microsoft Home, refer to the following URL: *www.microsoft.com/presspass/press/2006/sep06/09-28NewPrototypesPR.mspx*.

Microsoft is not the only company investigating the use of smart appliances. Government agencies and private companies have come to recognize the benefits of having devices that are able to react to their environments and make decisions. For example, the U.S. Department of Energy has been testing smart appliances such as thermostats, dryers, and water heaters that communicate with the power company via the Internet. By having the ability to intelligently adjust their power usage, these smart appliances have resulted in lower electricity bills for test participants.

Caring for the Elderly

According to the Centers for Disease Control (CDC), the average life expectancy for people in the United States has risen by more than two years since 1990. This trend has continued across most of the world and has dramatic implications for large populations such as in Japan. Before long, and in some places this is already the case, there will be more elderly people requiring assistance than there are workers able to care for them.

Robots will not replace human care workers, but they can help elderly people function safely within their own homes. For example, Japan has already developed an intelligent wheelchair prototype that can move around on its own and avoid obstacles. Rehabilitation robots that feature mechanical arms can be used to assist people in moving their limbs.

Universities have been working on projects to provide personal service robots to the elderly. Carnegie Mellon University (CMU) has developed a prototype robot named "Nursebot." The prototype robot the university has built allows care workers to remotely monitor in-home patients. It can also be used to remind patients when they need to take their medicine and directly interact with them to provide some level of companionship.

Private companies, such as PALS Robotics (see *www.palsrobotics.com*), produce prototype subsystems that can be used by researchers building home-care robots. For example, they have created an active stereo vision platform that can be attached to a mobile robot, allowing the robot to see.

Performing Dangerous Jobs

People have quickly realized that robots can be used to replace humans to perform dangerous jobs. For example, robots that disable bombs or explode land mines have been used for several years. These robots might even be suitable for use in uranium-processing plants, where the risk of exposure to radiation is very high. The trick is getting the robots to handle a wide range of possibilities. Although it is easy to have a robot perform a specific task under ideal conditions, it is not so easy to have a robot perform the same task when several dangerous and unexpected variables are at play.

In many cases, the Department of Defense has led efforts toward finding robotic solutions for dangerous jobs. In May 2007, Kansas State University received a $219,140 grant from the Department of Defense for building robots that serve as intelligent, mobile sensors. The robots Kansas State develops may be used to search buildings for weapons of mass destruction.

Also in 2007, the Defense Advanced Research Projects Agency (DARPA), which is the central research agency for the Department of Defense, sponsored the Urban Challenge (see *www.darpa.mil/grandchallenge/overview.asp*). This successor to the 2005 DARPA Grand Challenge involved autonomous vehicles operating in a simulated urban environment. The vehicles had to perform several tasks while also obeying traffic rules and avoiding other vehicles.

More than 100 university-based teams entered the challenge, but only 35 teams qualified for the semifinals by passing the rigorous on-site visits performed during the summer of 2007. Princeton University, one of the 35 teams to qualify, used a car powered by MSRS services (refer to the sidebar in Chapter 1, "Overview of Robotics and Microsoft Robotics Studio"). Unfortunately, Princeton University was not one of the 11 teams that made it to the finals. The finals, which were held in November 2007, resulted in only 5 teams crossing the finish line. The team from Carnegie Mellon University took first place, followed by the team from Stanford, and, in third place, the team from Virginia Tech.

The Urban Challenge was important because it represented an opportunity to explore applications for military and civilian uses. The Department of Defense hopes to use autonomous vehicles in dangerous battle situations. The driverless vehicles would be able to deliver needed supplies or rescue stranded troops. Private carmakers are hoping the innovations will lead to the design of intelligent, and thus safer, cars.

Performing Dull Jobs

For many years, robots have been used on assembly lines in manufacturing companies to perform dull and repetitive jobs. This is not difficult when the robot is bolted to the floor and asked to perform a very specific task. What is difficult is when the robot is mobile and able to encounter unexpected variables.

Domestic robots are now being used to perform boring household functions such as vacuuming the floors, cleaning the pools, and mowing the lawn. Imagine if the lawn-mowing robot encounters a baby bird or a child's toy on the lawn. Would it know not to run over the objects? Probably not. The area of robotic vision is one area still under development. It is one thing to recognize movement from Web camera images and another to identify and classify specific objects.

Unfortunately, we are a long way away from seeing a robotic butler, capable of operating in a house and doing our day-to-day chores. However, there are already robots designed to direct traffic or carry signs. As the field of robotics expands, we will see more and more robots performing jobs that humans would rather do without.

Performing Exploration Jobs

In 2004, NASA released the Personal Exploration Rover (PER) to a few museums throughout the United States. Museum visitors can use the Rover to explore an area meant to simulate Mars. The original Rover is still being used by NASA to explore the planet Mars. The Rover offers a relatively cheap and safe means of exploration (see *http://marsrovers.jpl.nasa.gov/home/index.html*). Rather than send astronauts to Mars, NASA sent two Rovers named Spirit and Opportunity. The Rovers were able to survive the severe dust storms on Mars and perform their missions year after year. Even though the Rovers were only supposed to operate on Mars for 90 days, they have performed so well that NASA continues to extend their missions, which may now last until 2009.

The experience gained from the Rover project is now being used toward a new project named the Telepresence Robot Kit (TeRK) (see *http://www.terk.ri.cmu.edu/*). TeRK extends upon the Rover museum installations and seeks to attract young men and women toward science and technology curricula.

Space is not the only frontier worthy of exploratory robots. Here on Earth, robots have been used to explore extreme temperature areas such as active volcanoes, desert environments, or regions in Antarctica. Carnegie Mellon University has created an army of these exploratory robots (see *www.frc.ri.cmu.edu/robots/*).

Additionally, autonomous underwater vehicles have been able to explore ocean depths inaccessible to humans. Robots such as the Autonomous Underwater Vehicle (AUV) from the University of Delaware (see *www.udel.edu/PR/UDaily/2005/mar/trembanis091405.html*) include learning capabilities that allow it to react to unexpected obstacles and chart new courses if necessary.

Providing Specialized Assistance

In the health care industry, robots are routinely used for performing surgeries and complex procedures. At this time, they function more as remote-controlled devices, but if significant advancements are made in the software used to control these devices, their uses could dramatically expand.

Since 2005, St. Mary's NHS Trust and Imperial College in London has used a remote-presence robot named Sister Mary (see *www.st-marys.nhs.uk/pressrd.html*) to roam the hospital wards. These robodocs are used by remote specialists to do virtual consultations with patients. The doctors can remotely operate the robot and send it from room to room. The patient is able to see an image of the doctor on the robot's attached computer screen. Sister Mary is not meant to replace doctors but rather to assist them in providing better health care for their patients.

Providing Companionship

Some robotics researchers predict that robotic companions will provide just as much benefit to people as live pets do. The outer shell for a companion robot can be as welcoming as a soft teddy bear. This is the case for the Huggable Robot (see *http://web.media.mit.edu/~wdstiehl/projects.htm*), which is currently under development at the Massachusetts Institute of Technology (MIT). The Huggable Robot can be used as an autonomous companion or as an avatar that allows people to remotely connect. MSRS was used for the avatar portion of this project.

Robotic companionship shows the most promise in the area of elderly care. Robotic companions can potentially provide a sense of comfort to an elderly person living alone or with little social contact. This is a concept known as animal-assisted therapy (AAT), and it is being used as a solution to the problems of loneliness and boredom.

Providing Entertainment

These days it is hard to walk into a toy store and not find several robotic toys. You will likely find one of the popular robotic toys from WowWee, such as the original Robosapien, or the new line of robotic dinosaurs. As the cost of mass producing these robotic toys decreases, you will see more and more children owning them.

Of course, robots used for entertainment are not restricted to toys for children (and even some adults). Using multiple robots, researchers and hobbyists have held competitions involving robots. For example, robot soccer is rapidly gaining popularity, along with the sumo-bot competitions, in which one robot tries to push another one out of a circle.

Integrating Artificial Intelligence

The field of AI is a vast field that covers a broad range of disciplines. There are numerous branches of AI that may be incorporated into robotics applications. Some of these branches are listed in Appendix A, "A Brief History of Artificial Intelligence." In many cases, the best scenario will be to combine multiple branches or techniques when designing a single robot application. This is something that Raúl Arrabales Moreno is doing with the robots he is designing (see the sidebar titled "Profile: Raúl Arrabales Moreno").

Profile: Raúl Arrabales Moreno

Raúl Arrabales Moreno is Web master for the Web site http://www.conscious-robots.com.

Raúl Arrabales Moreno hosts the Web site *www.conscious-robots.com*. The Web site is dedicated to scientific research in the area of machine consciousness and cognitive robotics. Arrabales, who is pursuing his PhD at the University Carlos III in Madrid, Spain, is also an instructor at the university. He supervises a group of graduate students who are interested in developing machine consciousness. Arrabales developed the Web site as part of his ongoing thesis work and had this to say about it:

> *I think we are currently living an exciting but still shy resurgence of the so-called Strong AI driven by the recent interest in the scientific study of consciousness. Over the last decade, significant advances and contributions from psychology, neuroscience, and philosophy have led part of the AI community to reconsider the possibility of engineering machine consciousness. The aim of the www.Conscious-Robots.com site is to contribute to the spread of the incipient field of machine consciousness and its potential application to robotics. I believe the scientific and engineering community involved in the design of the next generation of robots can greatly benefit from the sort of information and resources that are provided on the Web site. The site is also useful for students and researchers working in the field of robotics and using Robotics Studio.*

Arrabales has dedicated an entire portion of his Web site to MSRS (see *www.conscious-robots.com/en/robotics-studio/index.php*). It is here that Arrabales posts services he has developed with MSRS as part of his ongoing thesis research. He primarily works with the Mobile Robots Pioneer 3DX base. The project he is currently working on is named Conscious and Emotional Reasoning Architecture (CERA). The goal for this project is to develop and test various AI algorithms that result in automatic complex behavior generation. When asked what new feature he would like to see added to MSRS, Arrabales responded with the following:

> *From my point of view, one of the major advantages of using MSRS is the capability of easily managing concurrency and asynchronous input/output. However, a high-quality robotics application design requires that the correct coding patterns are used. I would love to see a tool for automatically generating widely used service coordination patterns. Maybe these patterns could come in the form of Visual Studio templates or a code generator like the one embedded in the Visual Programming Language. Using this feature, the user would select, for instance, to create a code template for writing a generic contract or to generate the required and optimal code for communicating with a GUI window object.*

Arrabales believes that the real challenge for robotics is in the field of autonomous mobile robotics. He foresees continued development in the field over the next 5 years but no significant advancements for at least 20 years. At that time, he hopes to see "a new generation of cognitive architectures that will be applied to robots, allowing them to learn new complex skills through interaction with other robots and humans."

Arrabales, who works primarily with MSRS and the Pioneer 3DX robot, thinks that "no single AI technique is good enough for the extremely complex needs of autonomous robot applications." He can envision scenarios where a genetic algorithm might be useful for generating new behaviors but then a neural network more appropriate for recognizing new faces. Arrabales does believe that the area of cognitive robotics offers the most promise.

Summary

- Like the field of AI, robotics has had its share of disappointments and failed predictions. This was partly due to limitations of the hardware, but it also is due to the complexity of the software needed.
- This chapter highlights potential applications for the field of robotics. These areas include smart appliances, caring for the elderly, performing dangerous or dull jobs, performing exploratory jobs, and providing specialized assistance, companionship, or entertainment.
- AI techniques such as genetic algorithms, machine learning, or neural networks can be used within robotics applications to make the robots more useful and able to react quickly to their environments.

Appendix A

A Brief History of Artificial Intelligence

The term artificial intelligence (AI) was first used in 1955 by a small group of researchers at Dartmouth College in Hanover, New Hampshire. The group started a summer project to determine if intelligence could be simulated using a machine. More than 50 years later, the world is still trying to make the same determination. In some ways we are closer to simulating human intelligence, and in other ways we are further away.

During the past 50 years, there have been more failures than successes in the area of AI. Although it seemed like a lot of progress was made in the late 1950s, the 1960s and 1970s forced researchers to realize that reproducing human intelligence was harder than they first imagined. The only applications that seemed to have much success with AI were games that involved a limited set of rules.

In the 1980s, with the introduction of the personal computer and increased computer usage, big companies used specific AI-based technologies such as machine learning and data mining to reduce costs and increase productivity. All of a sudden, researchers were able to deliver AI-based applications that justified the expense of creating them.

This trend continued into the 1990s, which saw significant advancements in computer capacity and the birth of the Internet. With this came the introduction of new concepts such as intelligent agents and natural interfaces. Other technologies that became popular included genetic programming and neural nets, which seemed to more closely mimic the way a human brain functions.

Over the years, a wide and ever-evolving range of technologies have been associated with AI. In some cases, a technology that becomes mainstream is no longer considered part of AI. This is mostly because AI-based technologies are seen as risky. This can make it confusing for developers trying to learn about AI. Table A-1 offers a list of branches or areas that are often associated with AI. Although this list does not include all areas, it should give you an idea of what techniques and technologies are currently considered branches of AI.

Table A-1 Branches of AI

Branch	Description
Agents	Involves autonomous software agents capable of representing and acting on behalf of their owners. To be considered intelligent, they should be able to make decisions by themselves based on criteria that have been predefined. Agents are able to communicate with people or other agents across distributed systems such as the Internet or a company network.

Table A-1 Branches of AI

Branch	Description
Game AI	The game industry is huge, and game developers often look toward AI to help them differentiate their products. By incorporating AI techniques such as machine learning, neural networks, and genetic algorithms, games can appear to be more interactive and tailored to the individual player.
Genetic Programming	This type of programming takes a cue from natural selection to determine the optimal solution. By creating and then evaluating several potential solutions, genetic algorithms are able to more accurately determine the best outcome. This type of programming can be very effective for scheduling or optimal path applications.
Intelligent Interfaces	This represents a wide range of natural interfaces that allow humans and computers to relate properly. Intelligent interfaces are software systems able to predict user behavior and anticipate user needs. These interfaces might need to incorporate other AI technologies to accomplish their goals. For example, the Bestcom application (out of Microsoft Research) can notify you of an incoming e-mail or call only when the sender is identified as important or the user is not busy. The application does this by utilizing techniques from areas such as machine learning, agents, and rule-based systems.
Machine Learning	Machine learning is often incorporated in a wide range of applications, such as data mining, speech recognition, natural language processing, and image processing. Several techniques, such as decision theory, neural networks, and genetic algorithms, may be utilized to effectively train a system to learn from example. The goal of machine learning is to have the computer learn from its experience and then adapt to the environment accordingly.
Recognition Tools	This involves the ability to recognize and identify an item or an object. This can include the recognition of voices through speech recognition and handwriting through optical character recognition (OCR). It can also include the recognition of a wide variety of objects through computer vision. In all of these areas, improvements in both the hardware and software needed to process data have allowed computer recognition tools to be used effectively in a wide variety of fields, such as robotics, security, manufacturing, farming, and health care.
Robotics	The robotics industry today features a wide variety of robots designed for specific purposes. The most effective robots perform tasks with a limited set of constraints and variables. These robots can be used for industrial and manufacturing purposes, as well as for social and domestic reasons. Many AI-based technologies, such as computer vision and machine learning, are used to enable robots to achieve their goals.

Where Is AI Today?

The past few years have brought about continued advancements in AI-based technologies. Most large banks use data mining and neural networks to search their databases for fraudulent transactions. Data mining involves the process of identifying patterns in a large set of data and, thereby, extracting meaningful information from that data. Neural networks are adaptive systems that use a network of interconnected artificial neurons to model complex relationships between inputs and outputs. This can be useful in detecting patterns and identifying relationships in data.

Many automakers are installing voice-recognition systems in their luxury cars. The popular OnStar service is now available in all General Motors cars. This service allows motorists to make hands-free phone calls, including calls for roadside assistance, using only their voices.

Speech recognition technology is commonly used by organizations to handle large volumes of inbound callers. Telephony-based applications have long been used by companies to route telephone calls, but advancements in speech recognition have allowed a broader range of companies to utilize telephony-based services. With increased accuracy and the availability of effective tools, companies can implement solutions that are natural-sounding and able to handle a variety of customer needs.

OCR, once considered a difficult task, is now available programmatically with Microsoft Office 2007. OCR, which is the process of translating written text into digital characters, is useful for tasks such as document imaging.

In the area of gaming, AI is particularly well suited because it operates with a limited set of rules and variables. In 1997, a chess-playing robot named Deep Blue beat grandmaster and defending world champion Garry Kasparov in a sensational chess match that lasted for days. Since then, there has been much debate about the progress of AI and whether a chess-playing robot should be considered intelligent. One thing is for sure, Deep Blue is essentially useless at doing anything other than playing chess.

The areas of AI that are the most difficult to tackle are those involving general-purpose tasks in an environment with unknown constraints. This is especially true when it comes to computer vision. Even though much progress has been made in this field, many more challenges lie ahead. For example, Stanford University recently won a prestigious award when it sent an autonomous vehicle across the Mojave Desert and won the DARPA Grand Challenge. Many lauded this success as a huge breakthrough in computer vision because this was one of the primary components needed to successfully steer the Stanford vehicle. Although the progress made by the Stanford group was significant, I doubt the vehicle used that year would have been successful if it was suddenly asked to navigate the busy streets of New York City instead.

Appendix B

Configuring Hardware

This appendix includes sections that cover how to configure the robotics hardware used in this book (the Parallax Boe-Bot, LEGO Mindstorms NXT, iRobot Create, and Arrick Robotics ARobot).

Configuring the Parallax Boe-Bot

The Parallax Boe-Bot robot (which is referenced in Chapter 3) processes operating instructions on a small stamp known as the BASIC stamp microcontroller. This single-board computer works fine for operating a small device such as the Boe-Bot, but it would be unable to accommodate even a compact version of the .NET Framework. For this reason, Microsoft Robotics Studio (MSRS) requires an interface that stands between the MSRS runtime and the computer that runs your robot.

The BASIC stamp runs the Parallax BASIC (PBASIC) language interpreter, which was developed by Parallax. Using the BASIC Stamp Editor, you can create programs that operate the Boe-Bot by issuing commands and interfacing with sensors. The latest version of the BASIC Stamp Editor is available as a free download from the Parallax Web site: *http://www.parallax.com/ProductInfo/Microcontrollers/BASICStampSoftware/tabid/441/Default.aspx.*

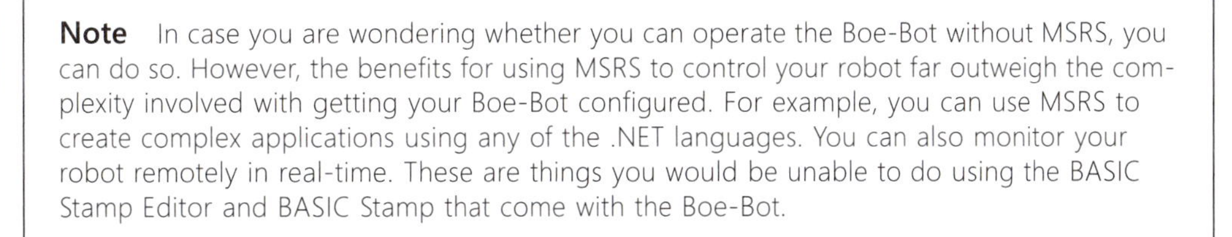

Note In case you are wondering whether you can operate the Boe-Bot without MSRS, you can do so. However, the benefits for using MSRS to control your robot far outweigh the complexity involved with getting your Boe-Bot configured. For example, you can use MSRS to create complex applications using any of the .NET languages. You can also monitor your robot remotely in real-time. These are things you would be unable to do using the BASIC Stamp Editor and BASIC Stamp that come with the Boe-Bot.

Install MSRS Interface

You need to download and install the BASIC Stamp Editor to configure your Boe-Bot to work with MSRS. This is the case regardless of whether you were using Visual Programming Language (VPL) or the Visual Studio template. You also need to download a zip file that contains PBASIC programs from the following URL: *http://www.parallax.com/Portals/0/Downloads/src/prod/MSRS-Bluetooth-Boe-Bot-Code-v1.2.zip*. You use the PBASIC programs to test your Boe-Bot and to enable it to work with MSRS. After you download from the Parallax Web site to your development machine, you need to use the BASIC Stamp Editor program to download the programs to the Boe-Bot. Specifically, you must download the

BoeBotControlForMsrsCtp2.bs2 program (see Figure B-1) to the robot using the serial or USB cable. The zip file from Parallax contains the following four PBASIC programs:

- **TestSpeakerLedsServos.bs2** This file should be downloaded to the Boe-Bot to ensure that the motors, speaker, and LEDs have been added correctly. After downloading, the Boe-Bot should beep, flash the lights on and off, and move forward, left, right, and then backward. If this does not happen, then you know there is a problem, and you should refer to the manual that accompanies the Boe-Bot.
- **TestWhiskersAndIr.bs2** This file needs to be used only if you have added the whiskers or infrared detectors to your Boe-Bot. The program displays the values for these sensors in the debug window, so you can verify that they are detecting objects correctly.
- **End.bs2** This file can be used before installing the eb500 Bluetooth module to ensure that electrical signals do not damage the wireless card.
- **BoeBotControlForMSRSCtp2.bs2** This file acts as a software driver for the Boe-Bot. Just like a printer driver allows your operating system and printer to communicate, this PBASIC program allows MSRS to communicate with your Boe-Bot. You must download this program to the Boe-Bot using the serial or USB cable before you can use VPL or any of the robotics tutorials.

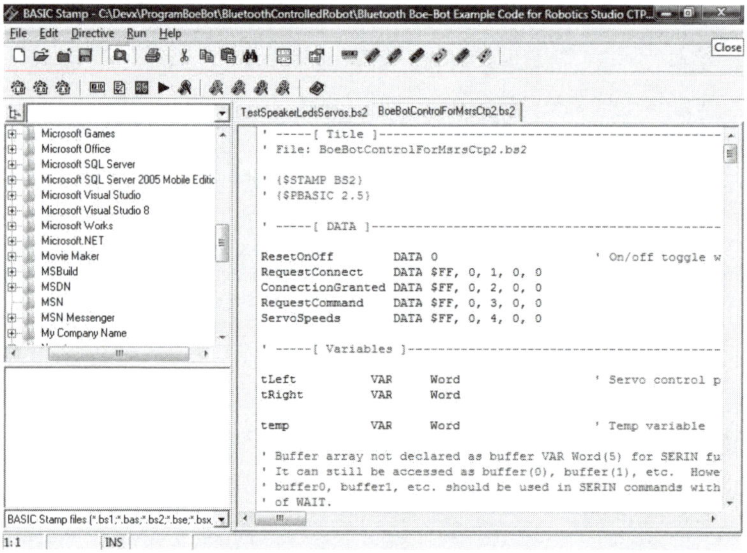

Figure B-1 The BoeBotControlForMSRSCtp2.bs2 PBASIC program must be downloaded to your Boe-Bot using the serial or USB connection before you can control the robot using MSRS.

> **Tip** The Boe-Bot comes in a serial and USB version. Check to see if your machine has a serial connection. If it does not, then you will need the USB version, which includes a USB to serial (RS232) converter and a USB A to mini B cable. You can combine these with your serial cable to connect the Boe-Bot without a serial connection on your computer.

Configure Bluetooth

The Boe-Bot comes with a serial or USB cable, which allows you to send commands to your Boe-Bot. If you use a cable to operate your robot, then it is considered tethered. This limits your ability to operate the robot because it can travel only as far as the three- or six-foot cable allows it to. Fortunately, the Boe-Bot provides support for a Bluetooth wireless connection through an eb500 Bluetooth module. The module comes with the Boe-Bot for the MSRS kit or is available for purchase separately from the Parallax Web site.

After installing the software that comes with your Bluetooth adapter, you will need to add the device using the Add Bluetooth Device Wizard. The Boe-Bot uses a passkey value of 0000. After entering the passkey, your operating system should assign the device to both an outgoing and incoming COM port. Record these numbers because you will need to enter them later.

MSRS needs to know what COM port to use when communicating with the Boe-Bot. In this case, it will be the *outgoing* port that will be referenced. To change the COM port reference, you will need to open the BASICStamp2 solution located in the MSRS installation directory under the \samples\Platforms\Parallax subdirectory. Open the file using Visual Studio 2005, and then open the Parallax.MotorIrBumper.config.xml file, which is located in the BSServices\Config folder. You need to change the element named SerialPort to contain the COM port used by your Bluetooth device. You also need to change the SerialPort element located in the Parallax.BoeBot.Config.xml file found in the \samples\Config\ subdirectory.

> **Tip** Andy Lindsay from Parallax maintains a document that details what you need to do to get the Boe-Bot to work with MSRS. This 20-page document lists more detail than is included here and should be referenced if you have any problems configuring your Boe-Bot. You can refer to the latest version of this document through the following URL: *http://www.parallax.com/dl/docs/prod/robo/MSRS-Bluetooth-Boe-Bot-v1.4.pdf.*

You then need to locate the private variable for the serial port, which is located in the BasicStamp2.cs file. Change the value for the variable named DefaultSerialPort to the value for your machine. The variable is located at the top of the class definition, such as in the following code:

```
public class BasicStamp2Service : DsspServiceBase
    {
        private const int DefaultSerialPort = 6;
```

Finally, you must build the BASICStamp2 solution and ensure that the build was successful. The solution includes two projects: BASICStamp2 and BSServices2. Changes are made to both projects, so you will need to recompile the entire solution.

> **Note** You should also pay special attention to whether your Bluetooth module is marked as eb500 SER C. You can find this on the module's serial tag. If it is, then you might encounter infrared (IR) interference that could cause your sensors not to perform properly. To avoid this, you need to request a new module or modify the existing module by snipping off one of the pin headers. Refer to the documentation listed in the previous tip for more information about this situation.

Enable Whiskers

By default, whiskers are disabled by the service that operates the Boe-Bot. If you decide to install the whiskers for your Boe-Bot, you must also make a change to the BoeBotControl.cs file in the BASICStamp2 project. The *ExecuteMain* method contains code that initiates a connection with the Boe-Bot and is responsible for enabling or disabling the appropriate sensors. To enable the Boe-Bot to use whiskers, you need to uncomment the call to *EnableWhiskers* and comment the call to *DisableWhiskers*, such as in the following code:

```
if (!handshake)
{
    handshake = WaitForConnect();
    irFlag = EnableIr();
    wFlag = EnableWhiskers();
    //wFlag = DisableWhiskers();
    digFlag = EnableDigitalSensors();
    for(int i = 0; i < 2; i++)
    {
        SetPins(3, 14, 3, 15);
        SpeakerTone(50, 4000);
        SetPins(4, 14, 4, 15);
    }
}
```

After you make the code changes, you must rebuild the entire BasicStamp2 solution and reload VPL before continuing.

Configuring the iRobot Create

Because the iRobot is preassembled, the steps required to set up this robot are less than what is required for the Boe-Bot. The Create, which we refer to in Chapter 5, comes with a serial cable. The serial cable is used to send commands from the computer to the robot. If your computer does not have a serial port connection, then you will need to purchase a USB to serial (RS232) adapter.

Tip It is not necessary to purchase the optional command package to work with MSRS.

Rather than alter the service code or XML-based configuration files, you can configure the COM port for your Create by using a configuration Web page. This configuration page is provided with the Create services, and it is initiated automatically the first time you try to use the Create with an MSRS service.

Upon first powering up the Create robot and attaching the serial cable between your computer and the Create, a COM port is assigned. If you are using Windows Vista, the port assignment appears in a status message box. You can also locate the port assignment by going to Device Manager. For Windows Vista users, you can do so by right-clicking Computer and selecting Properties. You can then click the Device Manager link and expand the Ports node (as shown in Figure B-2) to locate the port assignment.

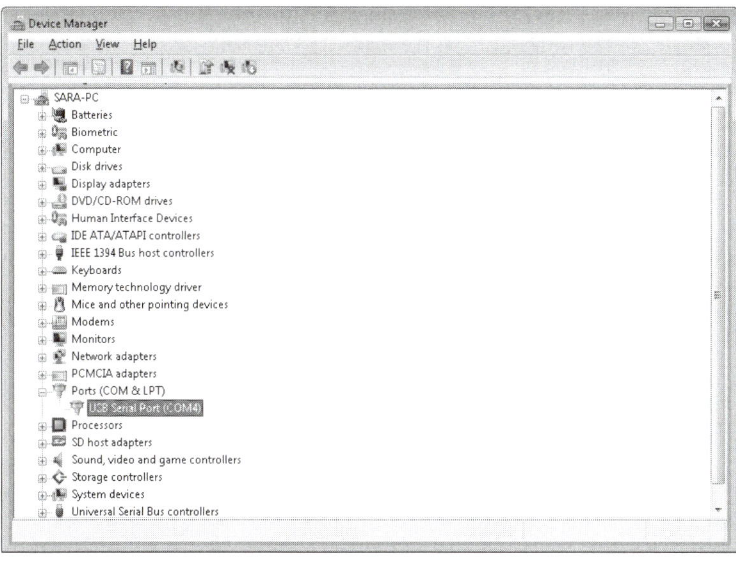

Figure B-2 You can use Device Manager to locate the port assignment for your Create robot.

The first time you execute an MSRS service that loads the Create manifest (such as the BasicDrive service provided with Chapter 5), the service determines whether the COM port has been configured. By default, the configuration file for the Create, which is physically located in the /samples/config folder for the local MSRS installation, contains no value for the SerialPort element. Rather than editing the XML-based config file directly, it is best to use the configuration page provided with the Create services.

To do this, copy the BasicDrive project from the book's companion Web site to a subfolder within the /samples folder for your local MSRS installation. Then, open the project using Visual Studio 2005. Go to the properties by selecting the Project menu and then choosing BasicDrive Properties. From the Debug tab, ensure that the path to the BasicDrive.manifest.xml file is correct. The path should be located in the Command Line Arguments text box. Also, ensure that the path to the iRobot.DriveBumper.manifest.xml file is correct. Start the project by clicking the green arrow button on the toolbar.

After the build is complete, you should see a Web page similar to Figure B-3. Enter the number assigned to the COM port in the Serial Port text box. Click the Connect button to connect to the Create robot. If the connection is successful, you should hear a short beep from the Create. It is possible for you to receive a message that says, "The iRobot is not connected. Please configure below." Ignore this message and start using the drive buttons on the BasicDrive windows form to operate the robot.

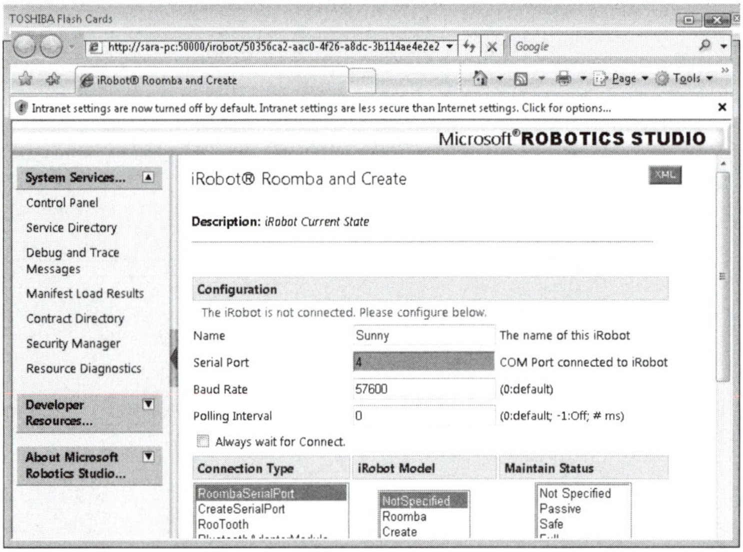

Figure B-3 Configuration Web page for the Create robot.

> **Note** You may have to enter valid credentials for your machine before connecting to the configuration Web page.

Configuring the LEGO Mindstorms NXT

When you first open the LEGO NXT box, you may feel overwhelmed by the number of LEGO pieces provided. Do not worry. It is not necessary for you to use all the pieces to begin working with MSRS and the LEGO NXT. You can simply use the Quick Start instructions to build the Tribot version of the LEGO NXT. You will then attach sensors, such as the sonar, touch, light, and sound.

Run the Installation

When you purchase a LEGO NXT, it should come with an installation CD that contains the software needed to operate your robot. The installation includes the LEGO Mindstorms NXT software version 1.0 and the LEGO Mindstorms NXT driver. You need to install this software on the development machine you use to communicate with the robot.

After the installation completes, use Windows Explorer to copy all files with a .ric file extension from the \samples\Platforms\LEGO\NXT\Resources folder for the MSRS installation to the \engine\Pictures folder for the LEGO installation. These are image files. They are needed because the LEGO program you will be loading is visual, and the images included with the MSRS installation are not included with the LEGO installation.

Configure Bluetooth

Because the LEGO NXT is already Bluetooth enabled, you do not need to install any additional hardware on the robot itself. However, you need a Bluetooth adapter on your development machine if it is not already Bluetooth enabled. After installing this adapter on your development machine, you must perform some additional steps to configure the LEGO NXT to work with Bluetooth.

The services provided with MSRS require that you use the Bluetooth connection, rather than a USB connection. This means that you will need to purchase a Bluetooth adapter if your computer is not already Bluetooth compatible. After installing the appropriate Bluetooth adapter, you can turn on the LEGO NXT and attempt to make a connection.

> **Tip** You need to ensure that the Bluetooth adapter you select is compatible with the LEGO NXT. The Bluetooth adapter that I used to communicate with the Boe-Bot (the Kensington, model #33348) did not recognize the LEGO NXT. I needed to purchase a D-Link, DBT-120 Bluetooth adapter to communicate with the LEGO NXT. Consult the following Web site for a list of NXT-compatible Bluetooth adapters, as reported by NXT users: *http://www.vialist.com/users/jgarbers/NXTBluetoothCompatibilityList*.

The top-left portion of the screen on the LEGO NXT brick contains information about your Bluetooth connection. If Bluetooth for the NXT is off, then no icon appears in this area. If there is a funny-looking, B-shaped symbol in the left corner, then the Bluetooth for the NXT is on, but your NXT is not visible to your Bluetooth adapter. If there is a funny-looking, B-shaped symbol and a left arrow (<), then your Bluetooth for the NXT is on, and the NXT should be visible to the Bluetooth adapter. Ensure that this symbol is present before continuing.

After you install the adapter and the NXT is accepting commands via Bluetooth, you can add the Bluetooth device. For Windows Vista users, you can do this with the Add Bluetooth Device Wizard. The wizard is accessible by right-clicking the Bluetooth icon (the funny-looking B shape) in your status toolbar and clicking Add Device. When the wizard appears, turn on the LEGO NXT and select the My Device Is Set Up And Ready To Be Found check box. Then, click Next.

The wizard searches for any available Bluetooth devices. If it finds the NXT, an NXT icon appears. Select this icon and click Next to continue. The next page asks you to provide a passkey. The LEGO NXT uses a passkey of 1234. To enter this, select the Use The Passkey Found In The Documentation radio button and type **1234** in the text box. Then, click Next. At this point, the display on the LEGO NXT should ask you to confirm the passkey. You do this by pressing the orange button on the LEGO brick.

At this point, Windows assigns a COM port to your Bluetooth device. This port assignment appears in a status box. You need to take note of the number for the outgoing COM port because you use it in the "Update Configuration Settings" section later in this appendix. Click Finish to complete the Add Bluetooth Device Wizard.

Install MSRS Interface

Like the Boe-Bot, the LEGO NXT requires an interface between MSRS and the actual robot. This interface runs on the NXT brick, which represents the brain for the LEGO NXT. The \samples\Platforms\LEGO\NXT\Resources folder for the MSRS installation includes a LEGO software program named msrs006.rbt. This is the program that acts as an interface between MSRS and the LEGO NXT. Double-click this file from Windows Explorer to load the program into the LEGO NXT development tool (see Figure B-4). It may take several seconds for the MSRS program to load, but you should wait until the program finishes loading before continuing to the next step.

When the program is visible in the development tool, you can click the arrow button in the bottom right-hand corner of the design surface. This downloads the program to the LEGO NXT. In order for this to work, the LEGO NXT has to be turned on and connected to your computer through a USB connection or the Bluetooth connection. Ensure that the download is complete before moving on to the next section.

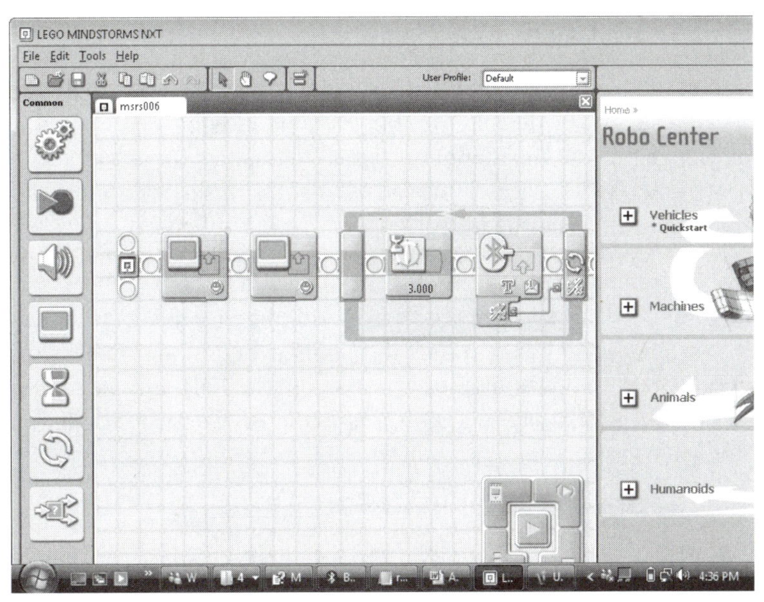

Figure B-4 Interface program, which allows MSRS to communicate with the LEGO NXT.

Update MSRS Services

In September 2007, the MSRS team released an updated set of services for the LEGO NXT. Not only do these services make it much easier to work with the LEGO NXT, they are also necessary to work with the code provided in Chapter 6. To get these services, you must download them from the MSRS Web site: go to *http://www.microsoft.com/downloads* and search on the words "samples update robotics."

You can update the NXT firmware by clicking Update NXT Firmware from the Tools menu in the LEGO NXT editor. You need a connection to the Internet because this step copies files to your LEGO NXT hardware. Do not assume that a recently purchased NXT has the latest firmware. Refer to the installation instructions included with the update, which, by default, are installed in the following location: C:\Microsoft Robotics Studio (1.5)\samples \SamplesUpdatePackage.htm.

Tip Make sure you use the USB cable provided with the NXT to download the firmware. If you try to download the firmware to the NXT using the Bluetooth connection, it will not be able to locate the device.

Update Configuration Settings

Like the Create robot, the services for the LEGO NXT include a Web page that makes configuring the services much easier. By default, the configuration files for the LEGO NXT do not include a COM port assignment. The number assigned to the outgoing Bluetooth port must be set before running any MSRS services with the LEGO NXT. This port assignment was made when you configured Bluetooth for the LEGO NXT.

The first time you execute an MSRS service using one of the LEGO NXT manifest files, you are directed to a configuration Web page (see Figure B-5). On this page, you must enter the COM port number in the available text box and click the Connect button to connect to the NXT. When the connection occurs, you should hear a beep from the NXT and see the words Microsoft Robotics Studio appear on the LEGO NXT brick. At this point, you have a successful connection, and your service should be able to execute.

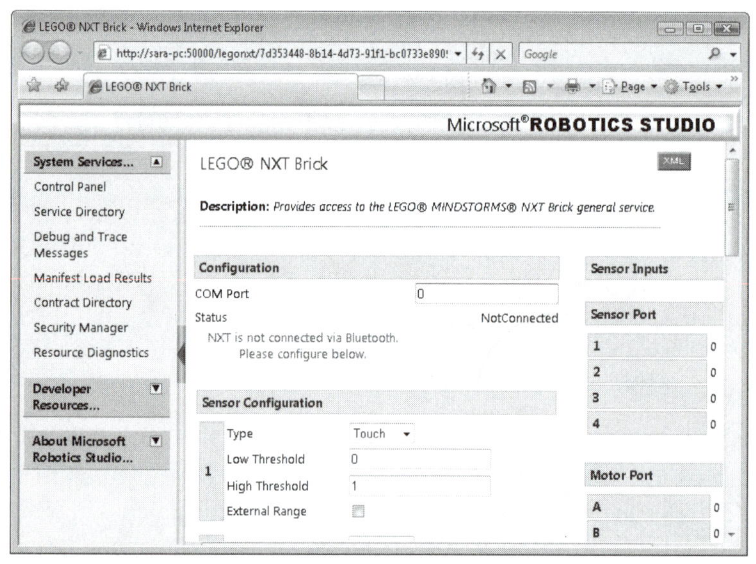

Figure B-5 Configuration Web page for the LEGO NXT robot.

Configuring the Arrick Robotics ARobot

Chapters 7 and 8 involve services that work with an unsupported robot named the ARobot by Arrick Robotics. The base model of this robot does not include Bluetooth connectivity, so you must use the supplied serial cable to connect your computer to the robot.

After you assemble it, you must follow the instructions provided in Chapter 7 to create a new hardware interface for this robot. This involves downloading a control program (available on the book's companion Web site) to the robot using the serial cable. You also need to build services used to control the ARobot and use them in a test program or in the SecurityMonitor project covered in Chapter 8.

Glossary

activity Basic building block used in a Visual Programming Language (VPL) application. The activity block can refer to a function such as calculate, or it can point to another service.

actuator Portion of a robot that is used to perform movement or accomplish a goal. For example, a robot's wheel is an actuator because it moves the robot. Additionally, a speaker would qualify as an actuator because it produces output.

analog sensor Device used to measure signals of varying strength, such as an audio signal.

asynchronous processing Considered an efficient means of processing multiple operations in parallel.

Bluetooth device Device used to receive a Bluetooth wireless connection. This type of connection is often used by robots to facilitate controlling the robot remotely.

binary large object (BLOB) Used to describe an object file that can store large amounts of unstructured data in a database management system.

Concurrency and Coordination Runtime (CCR) .NET library designed to reduce the amount of work involved in asynchronous programming. The CCR forms the base of the MSRS runtime engine.

contract Defines how a service can be used. It is used by other services, and HTML associated with the contract must be unique.

Decentralized Software Services (DSS) Lightweight, service-oriented runtime that sits on top of the CCR to form the MSRS runtime. The DSS combines Representational State Transfer (REST) principles with the architecture used to design Web services to deliver a powerful and flexible decentralized solution.

Decentralized Software Services Protocol (DSSP) Standard protocol designed by Henrik Nielsen and George Chrysanthakopoulos. The SOAP-based protocol extends upon the capabilities of standard HTTP operations to include support for subscriptions and event notifications.

deserialization Process of creating a .NET object based on the data obtained from a SOAP message.

differential drive system Used to identify a drive system that accepts alternating power levels to steer two wheels in any direction.

Direct Current (DC) Motor Conventional motor that is powered by a Direct Current and commonly found in robotics and toys.

DirectX Hardware Abstraction layer that is used to produce extensive visual three-dimensional effects. MSRS supports DirectX 9.

DSS Proxy The DSS proxy file is generated when the service is compiled and the assembly produced is used to represent a service.

dynamic-link library (DLL) Used to reference an external assembly; the DLL file contains a .dll file extension.

Entity The entity can represent any object within a simulation scene. This can include things such as the sky, the ground, or another robot.

Extensible Markup Language (XML) Designed by the World Wide Web Consortium (W3C), this standard represents a simple and flexible text-based format used to represent data.

Extensible Stylesheet Language Transformation (XSLT) For MSRS, the XSLT file is used to receive XML from a service and format the results as HTML or a text document, which can then be rendered in a Web browser.

Global Positioning System (GPS) Receiver Used to identify an object's location relative to feedback obtained from a satellite. The receiver can be used on a robot to precisely identify its location along a planned route.

Hypertext Transfer Protocol (HTTP) Communication protocol used to return formatted HTML that is displayed on a Web page.

infrared detector Device used to receive infrared signals, which can then be used to remotely control a robot.

Iron Python Version of the Python scripting language that Microsoft provides with the .NET Framework. MSRS developers may use the extensible and object-oriented language in their MSRS applications.

light-emitting diode (LED) Semiconductor diode that emits a light. Robots typically use a red or green LED to indicate something has taken place.

LEGO brick Brick-shaped object that contains the processor used by the LEGO NXT robot. It includes physical ports for each of the sensors included with the NXT.

manifest XML-based file that includes a list of services to be started. This is used by a robotics application to load any partner services.

MSRS service The core object in an MSRS application is a service. One or more Web-based services can be combined to form a single application. Each service will expose state, which is used to define the service. The service can be thought of as an active document that exposes information.

partner Represents a service that is tied to another service. This is done to expose the state of one service and establish a relationship between services.

port For MSRS, the port represents a way for services to communicate with each other. Messages are sent back and forth through the port.

Representational State Transfer (REST) Introduced by Roy Fielding, REST encompasses a set of principles used to define the technologies for the World Wide Web. The REST architecture was incorporated into the design of the MSRS runtime.

sensor Used to gather input about a robot's environment. For example, a contact sensor is used to determine when the robot has encountered an object. Most robots use at least one sensor to gather information.

serialization Process of representing a .NET object as a stream of bytes. This is typically done to produce a SOAP message.

servo motor Motor in which the entire system is contained within a small plastic block.

SOAP Designed by the World Wide Web Consortium (W3C), this transport protocol provides a standard way to send messages between applications using XML.

state Used to represent a service, the service can expose one or more items through the state. The state represents the service at the time the service is requested.

transform An assembly that acts as a bridge between the service and the proxy assembly by providing a mapping between the service and the proxy.

Transmission Control Protocol (TCP) Standard protocol used to transport data efficiently across a network.

Visual Programming Language (VPL) Graphical language tool that is provided with MSRS. VPL allows developers to build simple or complex robotics applications by dragging and dropping blocks onto a design surface.

Visual Simulation Environment (VSE) Rendering tool available in MSRS that allows you to modify simulation scenario scenes and execute simulations. VSE is powered by the AGEIA PhysX engine, which allows it to render advanced physics graphics.

XML (Extensible Markup Language) Designed by the World Wide Web Consortium (W3C), this standard represents a simple and flexible text-based format used to represent data.

XNA XNA, which is not acronymed, is a layer that sits on top of DirectX to allow for three-dimensional rendering. This is used by the simulation environment in MSRS.

XSLT (Extensible Stylesheet Language Transformation) For MSRS, the XSLT file is used to receive XML from a service and format the results as HTML or a text document, which can then be rendered in a Web browser.

World Wide Web Consortium (W3C) This independent standards body is responsible for creating and publishing standards used to operate the World Wide Web.

Bibliography

Anthes, Gary. "Self-Taught: Software that Learns by Doing." *Computerworld*, February 6, 2006. *http://www.computerworld.com/softwaretopics/software/story/0,10801,108320,00.html.*

Gates, Bill. "A Robot in Every Home." *Scientific American*, January 2007. *http://www.sciam.com/article.cfm?chanID=sa006&articleID=9312A198-E7F2-99DF-31DA639D6C4BA567.*

Hall, Michelle. "K-State Professor Works to Create Cooperative Robot Teams to Handle Hazardous Jobs." *Kansas State Press Release*, May 1, 2007. *http://www.k-state.edu/media/newsreleases/may07/deloach50107.html.*

International Federation of Robotics. "2005 World Robot Market." 2005. *http://www.ifr.org/statistics/keyData2005.htm.*

Keizer, Gregg. "Microsoft Predicts the Future with Vista's SuperFetch." *InformationWeek*, January 19, 2007. *http://www.informationweek.com/news/showArticle.jhtml?articleID=196902178.*

Kitano, Massayuki. "Japan Looks to Robots for Elderly Care." *Fairfax Digital*, July 22, 2005. *http://www.theage.com.au/news/technology/japan-looks-to-robots-for-elderly-care/2005/07/21/1121539082236.html.*

Lai, Eric. "Microsoft: Future Homes to Use Smart Appliances, Interactive Wallpaper." *Computerworld*, September 29, 2006. *http://www.computerworld.com/action/article.do?command=viewArticleBasic&articleId=9003752.*

NASA. "NASA Extends Operations for Its Long-Lived Mars Rovers." *NASA Press Release*, October 15, 2007. *http://marsrovers.jpl.nasa.gov/newsroom/pressreleases/20071015a.html.*

Nielsen, Henrik, and George Chrysanthakopoulos. "Decentralized Software Services Protocol–DSSP." November 6, 2006. *http://download.microsoft.com/download/5/6/B/56B49917-65E8-494A-BB8C-3D49850DAAC1/DSSP.pdf.*

Orenstein, David. "New Algorithm Improves Robot Vision." *Stanford Report*, December 7, 2005. *http://news-service.stanford.edu/news/2005/december7/robotsee-120705.html.*

Richter, Jeffrey. "Concurrent Affairs: Concurrency and Coordination." *MSDN Magazine*, September 2006. *http://msdn.microsoft.com/msdnmag/issues/06/09/ConcurrentAffairs/default.aspx.*

Roy, Nicholas, Gregory Baltus, Dieter Fox et al. "Towards Personal Service Robots for the Elderly." Carnegie Mellon University, 2000. *http://web.mit.edu/nickroy/www/papers/wire2000.pdf.*

Sedecca, Boris. "Best Kept Computer Secrets Revealed." *ComputerWeekly*, October 12, 2006. *http://www.computerweekly.com/Articles/2006/10/12/219087/best-kept-secret-agent-revealed.htm.*

Skillings, Jonathan. "FAQ: Keeping Pace with Robots" *CNET News*, October 5, 2005. *http://news.com.com/FAQ+Keeping+pace+with+robots/2100-11394_3-5889478.html.*

Stiehl, Dan, Jeff Lieberman, Cynthia Breazeal et al. "The Huggable: A Therapeutic Robotic Companion for Relational, Affective Touch." *IEEE Consumer Communications and Networking Conference*, January 2006. *http://web.media.mit.edu/~wdstiehl/Publications/StiehlHuggableCCNC06FinalInfo.pdf.*

U.S. Centers for Disease Control. "Table 27: Life Expectancy at Birth." 2006. *http://www.cdc.gov/nchs/data/hus/hus06.pdf#027.*

Williams, Martyn. "Robots Take Dangerous Jobs." *PC World*, April 3, 2003. *http://www.pcworld.com/article/id,110127-page,1/article.html.*

Woodall, Bernie. "Smart Appliances Have Minds of Their Own." *Reuters*, January 19, 2007. *http://www.boston.com/business/technology/articles/2007/01/19/smart_appliances_have_minds_of_their_own.*

Index

Symbols and Numbers

A

B

C

D

E

F

G

J

K

L

M

N

O

P

S

T

U

V

W

X

Y

Z

About the Author

Sara Morgan is an independent developer and author who writes articles and books concerning key Microsoft technologies. Sara operates her own consulting firm named Custom Solutions, LLC. An MCSD and MCDBA, she specializes in niche areas such as speech technologies and robotics. She has extensive experience with SQL Server and Visual Studio .NET and has been developing database-driven Web applications since the earliest days of Internet development.

For the past two years, her main client has been Microsoft, and, during that time, she has co-written four training kits for Microsoft Press. These training kits are used by developers to study for certification exams, and they cover topics such as distributed development, Web application development, SQL Server query optimization, and SQL Server Business Intelligence. Sara has also written several articles for the online development journal DevX.com concerning Microsoft Speech Server and the newly released Microsoft Robotics Studio (MSRS). In early 2007, she was named a Microsoft MVP for the Office Communications Server (OCS) group.

What do you think of this book?

We want to hear from you!

Do you have a few minutes to participate in a brief online survey?

Microsoft is interested in hearing your feedback so we can continually improve our books and learning resources for you.

To participate in our survey, please visit:

www.microsoft.com/learning/booksurvey/

...and enter this book's ISBN-10 or ISBN-13 number (located above barcode on back cover*). As a thank-you to survey participants in the United States and Canada, each month we'll randomly select five respondents to win one of five $100 gift certificates from a leading online merchant. At the conclusion of the survey, you can enter the drawing by providing your e-mail address, which will be used for prize notification only.

Thanks in advance for your input. Your opinion counts!

*Where to find the ISBN on back cover

Example only. Each book has unique ISBN.